REDUCING THE IMPACTS OF DEVELOPMENT ON WILDLIFE

James Gleeson
and Deborah Gleeson

CSIRO
PUBLISHING

National Library of Australia Cataloguing-in-Publication entry

Gleeson, James.

Reducing the impacts of development on wildlife/by James Gleeson and Deborah Gleeson.

9780643100329 (pbk.)
9780643106932 (epdf)
9780643106949 (epub)

Includes bibliographical references and index.

Sustainable development.
Wildlife conservation.
Habitat (ecology).
Environmental policy.
Natural resources – Management.
Habitat conservation.

Gleeson, Deborah.

338.927

Published by
CSIRO PUBLISHING
150 Oxford Street (PO Box 1139)
Collingwood VIC 3066
Australia

Telephone: +61 3 9662 7666
Local call: 1300 788 000 (Australia only)
Fax: +61 3 9662 7555
Email: publishing.sales@csiro.au
Web site: www.publish.csiro.au

Front cover: Australian pelican (photo: Ron Whitney), city skyline (photo: iStockphoto)

Set in 10/12 Adobe Minion Pro and ITC Stone Sans
Edited by Peter Storer Editorial Services
Cover and text design by James Kelly
Typeset by Desktop Concepts Pty Ltd, Melbourne
Printed in China by 1010 Printing International Ltd

CSIRO PUBLISHING publishes and distributes scientific, technical and health science books, magazines and journals from Australia to a worldwide audience and conducts these activities autonomously from the research activities of the Commonwealth Scientific and Industrial Research Organisation (CSIRO). The views expressed in this publication are those of the author(s) and do not necessarily represent those of, and should not be attributed to, the publisher or CSIRO.

Original print edition:
The paper this book is printed on is in accordance with the rules of the Forest Stewardship Council˚. The FSC˚ promotes environmentally responsible, socially beneficial and economically viable management of the world's forests.

MIX
Paper from
responsible sources
FSC® C016973

Dedication

To our grandparents and parents who
fostered our keen interest in animals and plants
during our childhood.

Foreword

I have been fortunate to work across the disciplinary boundaries of environmental science, ecological research and consulting practice for most of my 35-year career. After 6 years working in environmental chemistry in research organisations in Australia and the USA and 13 years with the international consulting company, Dames & Moore, I joined CSIRO Wildlife and Ecology in 1995. My role was to help the organisation with the process of brokering ecological research into practical applications of knowledge for improved wildlife conservation and natural resource management on development projects. Throughout my career, I have been involved in the environmental assessment of many major infrastructure projects for the mining, oil and gas industries in remote regions, and for urban and industrial developments in cities. And now I found myself working with some of the best wildlife researchers in Australia located at Gungahlin Homestead in Canberra.

My experience as an environmental scientist working on controversial infrastructure projects inevitably led me to working with people who brought widely different knowledge and perspectives on the value of wildlife. For example, I have listened carefully to ecologists, natural historians, traditional owners, land managers, community-based organisations, regulators, engineers, lawyers, developers and passionate citizens speaking out about wildlife – and seldom with a shared view.

In trying to resolve these contending rationalities, there is always some common disputed ground between the economic values of a developable resource and the biodiversity values of viable wildlife and functional ecosystems – as if there were a single choice between development or conservation. And more often than not, either an iconic species of Australian wildlife, or one that became well known through the project, has found itself close to the centre of an increasingly politicised decision-making process – sometimes landing wildlife on the front pages of newspapers or even in the courts.

In the 1970s and 1980s, with the rise of environmental impact assessment and a resource-driven economy, there were many field surveys conducted on sites that turned up new information about wildlife species for particular projects. Most of these local surveys for companies were conducted by consultants and the very few big regional fauna surveys were conducted by CSIRO, state departments and some universities. However, at that time, even basic, reliable published information about wildlife distribution was not as easy to access as it is nowadays. And published ecological research almost always sat in hard copy journals in research organisations and university libraries. This meant that there was an asymmetry of information access between what was available to the researchers and what was accessible for those in consulting and management practice who were often living in remote areas without access to any libraries.

In the 1980s, most environmental management staff and consultants working on major infrastructure projects went about their environmental compliance requirements without much deliberate focus on wildlife, except to avoid accidental loss or to conduct a routine fauna survey. The focus was mostly on the rehabilitation of degraded landscapes and water management – gaining practical knowledge along the way from their own experiences and other knowledge from that of the traditional owners, regulators, occasional researchers and local land users who visited the development sites. In general, this practical ecological knowledge was only written up in the grey literature of conference proceedings and shared by word of mouth through industry-sponsored environmental conferences or by a consultant transferring the knowledge by working on another project.

How the game has changed in the last two decades! The 1990s and 2000s saw significant advances in funding large-scale research and development programs for fauna conservation. For example, Western Shield in Western Australia was launched in 1996 and is a world-class predator control program aimed at the recovery of a range of threatened species, mostly through the wise use of 1080 against foxes. There were also three consecutive Cooperative Research Centres (CRC) – namely Vertebrate Biocontrol, Pest Animal Control, and Invasive Animals that brought together traditional strengths in wildlife research in CSIRO and the universities together with state departments and biocontrol organisations to consider new control measures for major invasive species such as foxes, rabbits, mice, cane toads and carp.

There were also large-scale vegetation and fauna surveys conducted as part of the Regional Forest Agreements and the State Forests of NSW Environmental Impact Assessments in the 1990s. These initiatives required extensive field surveys for fauna using spatial analysis of vegetation patterns for habitat assessment and display of information within a new range of accessible geographic information systems. These techniques were complemented by the development of ecological theory around population viability, database analysis and adaptive management approaches.

It was only a matter of another few years before the rise of the internet and easy access to powerful search engines. These technologies have revolutionised the information landscape for land managers, consultants, researchers and community-based organisations, with unparalleled access to electronic environmental information in the form of satellite imagery, peer-reviewed journals, data transfer, remote sensing and monitoring technologies. All of a sudden, there became a form of electronic accountability whereby information is able to be easily accessed, evaluated and scrutinised by many other stakeholders entering the development process – and if not, why not?

When I was first approached to provide a foreword for *Reducing the Impacts of Development on Wildlife* I did not realise that I would be reading a book that overcomes the information asymmetry problem we experienced in wildlife research and management in the last century. This book, written by ecologists with backgrounds in both research and consulting practice, provides an excellent synthesis of the peer-reviewed research literature across many disciplines. However, the real advance has been to combine ecological theory and field research with practical knowledge from many local case studies in wildlife management across many development projects.

I do not recall seeing a book before that so clearly outlines in clear technical English so many practical measures for reducing impacts on wildlife across multiple types of industrial and urban developments. This book will undoubtedly fill an empty niche on the bookshelves of wildlife practitioners, managers, researchers, students and policy makers. The authors also invite us to join a process for rapid learning and knowledge sharing among all those who value and want to protect our wildlife. Let's join them!

Allen Kearns
CSIRO Ecosystem Sciences

Contents

Acknowledgements

We would like to thank Allen Kearns for contributing the foreword as well as Darryl Jones, Andrew Hamer, Aaron Organ and David Goldney for contributing case studies.

We thank those who reviewed the draft chapters: Martin Denny, Allen Kearns, Darryl Jones, Ross Goldingay, David Goldney, Kyle Armstrong, Robert Bullen, Glenn Muir, Arthur White, Michelle Gleeson, Danielle Gleeson, Sean McNamara, Ben Smith and Kerrie Jocumsen-Smith. Their contributions have improved the book. Norrie Sanders kindly shared his ideas in the early stages of this book.

We are grateful to Julia Toich who assisted with preparation of the schematics and the people who assisted with photographs: Ron Whitney, Rodney van der Ree, Stephen White, Paul Smith, Catherine Offord, John Koch and Michelle Gleeson, among others. We would also like to thank the CSIRO Publishing team.

Lastly, but not least, we would like to thank our families and friends for their support while being consumed writing this book. We apologise to anyone else we may have overlooked.

About the authors

James Gleeson
James Gleeson has a background in ecological research, consulting and management. He has spent the past decade working for some of the largest mining and infrastructure projects around Australia. James works alongside some of Australia's leading botanists, zoologists and ecologists to develop measures to overcome disparities between development and conservation of biodiversity. James received his Bachelor of Science with Honours from the University of Queensland.

Deborah Gleeson
Deborah Gleeson has specialised in fauna and flora for over a decade, during which time she gained a PhD in ecology and evolutionary biology, and worked as a field ecologist and ecological project manager for environmental consulting companies, among other roles. Deborah has played a key role in recommending measures to reduce impacts on flora and fauna for a variety of residential, commercial, industrial, infrastructure and mining developments. Deborah is an Honorary Fellow at Griffith University School of Environment.

Contributors

Darryl Jones

Associate Professor Darryl Jones is the Deputy Director of the Environmental Futures Centre at Griffith University. Darryl directs the Applied Road Ecology Group, where he is currently working in partnership with the Queensland Department of Transport and Main Roads on a major project aiming to reduce koala-related vehicle mortality. He has co-authored several publications on efficacy of the first purpose-built fauna crossing structures in Australia, as well as other aspects of human–wildlife interactions.

David Goldney

Professor David Goldney is an Adjunct Professor at Charles Sturt University and the University of Sydney, as well as the Principal Consulting Ecologist for Cenwest Environmental Services. David has undertaken fauna assessments for numerous mining projects. He also held the position of Acting Commissioner of the New South Wales Land and Environment Court.

Allen Kearns

Mr Allen Kearns has been with CSIRO since 1995, where he has held the position of Deputy Chief of Sustainable Ecosystems, among other roles. He previously worked for 13 years with Dames & Moore as an environmental scientist in Australia, the USA and France. He has held positions, including Chair, on several independent panels of experts for proposed developments.

Andrew Hamer

Dr Andrew Hamer is an ecologist at the Australian Research Centre for Urban Ecology (ARCUE) at the Royal Botanic Gardens Melbourne. He has published extensively on frog ecology and has 12 years experience as an environmental consultant. His PhD research focused on the ecology of the threatened green and golden bell frog *Litoria aurea* in New South Wales.

Aaron Organ

Mr Aaron Organ is the Director/Principal Ecologist at Ecology and Heritage Partners. He has over 16 years experience in the environmental field and has provided ecological advice regarding proposed residential and industrial subdivisions, and large infrastructure projects such as wind farms throughout south-eastern Australia. He has also authored or co-authored over 50 ecological reports, research papers, and management plans and strategies relating to the ecology and conservation of the nationally threatened southern bell frog *Litoria raniformis*.

1

Introduction

Why this book was written

It seems that everywhere you look in Australia right now there is some sort of development going on. Like many countries around the world, cities and towns are ever stretching across once natural landscapes, transport networks are spreading and natural resources are being consumed faster than ever. The high rate of development in Australia is partly fuelled by its thriving resources sector servicing unprecedented overseas demand for minerals (particularly coal and iron ore). New mines are springing up, established mines are expanding and rail and port infrastructure is being built to overcome transport bottlenecks. At the same time, sharp population increases have taken place. A staggering 421 300 people were added to Australia's population in 2009 alone, with the population predicted to possibly almost double to 42.5 million by 2056 (Australian Bureau of Statistics 2008; Australian Bureau of Statistics 2011a). In recent years, Australia has had the fastest-growing population in the developed world, easily outpacing the growth rates of China and India (Smith 2011). More people results in a need for more residential and commercial developments, as well as supporting infrastructure such as powerlines, roads, pipelines and bridges.

Developments are, by and large, encouraged by governments around the world because they bring affluence to a country in the form of jobs and a stronger economy. However, more development places greater pressure on the natural environment. Some developments have a greater impact on the environment than others, depending on their location and size as well as how they are designed and managed. The consequences of poor planning of development may include permanent loss or degradation of important flora and fauna habitat, disruption of dispersal pathways and extinction of species.

In the pursuit of preventing adverse impacts on the environment, there is a push for development to be 'sustainable', in preference to curbing growth or the factors driving it. 'Ecologically sustainable development' is a policy concept that involves undertaking development to meet human needs in such a way as to conserve and/or enhance natural ecosystems (Ecologically Sustainable Development Steering Committee 1992; Hezri and Dovers 2009).

Overall, biodiversity continues to substantially decline in many parts of Australia (Beeton *et al.* 2006). Now, more than ever, it is crucial that we ensure that developments are undertaken in a manner that reduces impacts on the environment and that the phrase 'ecologically sustainable development' is not just used as a trendy catchcry to justify development. In particular, reducing impacts of development on native wildlife (plants and animals) is essential for the maintenance of biodiversity and preservation of functioning ecosystems.

There is reason for hope and optimism because measures to reduce impacts of development on flora and fauna are increasingly being incorporated into developments in Australia and many other parts of the world. Practical measures must be correctly selected, designed and implemented not only to ensure the best outcome for wildlife but also to avoid unnecessary costs to enable the development to run productively. Development and conservation are not necessarily mutually exclusive. Rather, both can be achieved where there is genuine commitment to conservation.

So, how effective are the measures currently used to reduce the impacts of development on wildlife? In recent years, many of these measures have been adopted as industry standards without rigorous scientific testing to validate them. Other measures have been tested, but the results of this testing often remain hidden in scientific journals that are not readily accessible to the general public, instead of influencing how these measures are applied. If a development relies upon a particular measure to reduce impacts on wildlife, the design, effectiveness and reliability of that measure must be well established: otherwise it risks failing. In some situations, incorrect application of a measure can also be damaging to the environment. These problems make it difficult for those involved to implement the best measures to minimise the impacts of development on wildlife.

In Australia, the requirement for developers to implement measures to reduce the impacts of development on wildlife is underpinned by government legislation and the environmental impact assessment process for new developments. Developers usually turn to environmental professionals for advice on how to meet their obligations. These professionals are now more than ever responsible for providing sound advice, so it is crucial that they have the most reliable information at their fingertips. We wrote this book so that anyone involved in reducing the impacts of development on wildlife could have easy access to current information. This book has been written with the hope that measures are applied more consistently for similar impacts on wildlife across different types of development and across different states and territories in Australia. We believe this book to be the first of its kind in Australia.

The scope of this book

The overall aim of this book is to describe and evaluate the effectiveness of key measures used to reduce the impacts of development on flora and fauna. The measures described in this book have been designed to reduce the impacts arising from various types of developments including residential, commercial, industrial, infrastructure, mining, agriculture and forestry. Naturally, it is not feasible to write a book that includes *all* of the measures to reduce *all* types of impacts of developments on *all* types of flora and fauna. To do this would be a mammoth, if not impossible, task. Instead, we have chosen to narrow our focus as follows.

We have taken an Australian perspective when writing this book because the Australian flora, fauna and environment is unique in many ways and often dictates the design of measures used to reduce impacts. Many of the measures described in this book have been used somewhere in Australia at one time or another. To add to this, we have investigated whether other measures used successfully overseas could be adapted for use in Australia. Despite our Australian perspective, this book is likely to be useful to readers in other countries. Overseas readers may find that some of the measures uniquely used in Australia, or their variations, afford suitable solutions for their particular problems.

The term 'wildlife' is used in this book to describe terrestrial flora and fauna. This book does not cover the impacts of development on flora and fauna in marine and freshwater ecosystems. While undoubtedly important, aquatic flora and fauna deserve proper treatment in separate dedicated books.

Just as developments can adversely affect wildlife, sometimes wildlife can cause problems for people and their developments – the other side of the human–wildlife conflict. We

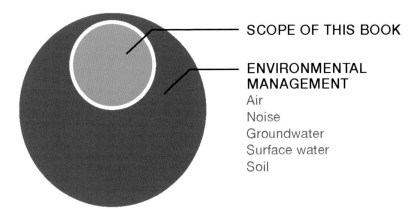

Figure 1.1. The scope of the book in relation to the larger topic of environmental management (Diagram: J Toich)

emphasise here that this book does not cover measures that have been developed to minimise the impacts of *wildlife* on *development*. For example, measures to prevent flying foxes from foraging on horticulture crops or native birds from perching on lamp posts are not covered.

The scope of this book focuses on management of the biotic environment. We recognise that physical changes to the abiotic environment (e.g. soil, air and water) can also impact wildlife and management of these components plays a crucial role in reducing the various impacts of development on wildlife (Figure 1.1). For example, changes to groundwater level and/or quality can affect ecosystems dependent on groundwater. In these situations, it is more efficient to reduce impacts on wildlife by applying measures that address the impact on the groundwater.

We recognise that environmental policy can be a powerful means for ensuring that impacts of development on wildlife are minimised (e.g. the Queensland *Vegetation Management and Other Legislation Amendment Act, 2004* which has restricted land clearing in Queensland); however, discussion on environmental policy is outside the scope of this book. This book does provide information that can re-focus expectations concerning the application of existing policies, such as ecologically sustainable development or perhaps the creation of new policies.

The effectiveness of every measure included in this book has been critically analysed by us. We found some measures to be highly effective in reducing impacts on wildlife, others to be lacking and, for some measures, there has not been enough testing to make a judgement either way. Our conclusions are supported by references drawn from a variety of sources including scientific literature as well as personal experience. Some industry myths associated with the use and effectiveness of measures are also discussed.

The large amount of information relevant to the topic of this book can be somewhat over-whelming. One of our aims was to interpret and consolidate this information to help bridge the gap between scientific research and practical application. Nevertheless, tailored advice is important when aiming to reduce the impacts of a particular development on wildlife (see Box 1.1). We have included case studies in this book from prominent scientists and practitioners from Australia.

This book was written for anyone involved in reducing the impacts of development on wildlife. We anticipate a wide audience, including those working in the private sector of the environmental industry (e.g. environmental consultants and managers), developers (e.g. residential, infrastructure, industrial and mining), various levels of government (e.g. environmental officers, strategists, policy makers and government regulators) as well as community

Box 1.1   **Why tailored advice is important when aiming to reduce the impacts of a particular development on wildlife**

Each development scenario is unique and the potential outcomes of using (or not using) a measure to reduce impacts on wildlife can vary widely, depending on the particularities of the development as well as the nature of the receiving environment. Some measures intended to reduce impacts on wildlife can, in fact, harm the environment in some circumstances. Some measures may be effective at one development site but not at another.

This is why it is important to consult a qualified and experienced ecologist for each individual development scenario. It is necessary to ensure that the ecologist consulted is familiar with the specific development and its environment so they can determine the most appropriate way to reduce its impacts on wildlife while minimising harm to other aspects of the environment.

This book contains general information only and is not a substitute for tailored advice from a qualified and experienced ecologist. No reader should act on the basis of any matter contained in this book without first seeking appropriate professional advice that takes into account their own particular circumstances.

environmental groups and students. This book may also help researches identify where additional research could benefit wildlife management in Australia.

We have written this book to provide a framework for reducing the impacts of development on wildlife. Importantly, we hope that this book will energise the topic and inspire others to come up with their own innovative ideas. There can be little doubt that there are more measures to reduce impacts on wildlife still waiting to be invented.

The structure of this book

We have structured this book so that it can be read from cover to cover as well as used as a reference manual as specific solutions are needed. We discuss the measures to reduce the impacts of development on wildlife assuming that the reader has a general knowledge of the impacts of common developments and an understanding of ecology. However, those readers who could

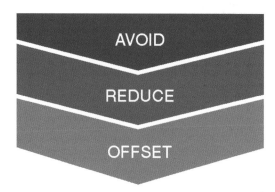

Figure 1.2.   The environmental impact assessment hierarchy (Diagram: J Toich)

Table 1.1.   Function of measures included in Chapters 5 to 11

Chapter	Function of measure
5	Modify human behaviour to reduce impacts on wildlife.
6	Reduce direct wildlife mortality during land clearance.
7	Exclude or deter fauna from a development site.
8	Promote safe movement of fauna around and through a development site.
9	Reduce habitat degradation near a development site.
10	Reduce impacts on wildlife displaced from a development site.
11	Enhance flora and fauna conservation *ex-situ*.

benefit from some basic background information will find an outline of some potential impacts in Chapter 2 and an overview of some commonly discussed ecological concepts in Chapter 3.

The structure of the remaining chapters broadly follows the environmental impact assessment hierarchy (Figure 1.2). According to this hierarchy, potential impacts on biodiversity are first and foremost avoided. When impact avoidance is not feasible, priority is given to mitigating (i.e. reducing impacts). Any impacts that cannot be mitigated are offset (i.e. compensated).

Impacts on wildlife can often be avoided by simply moving the development to an alternative location or by modifying the layout of a development at the planning stage. However, as discussed in Chapter 4, there are many more ways to avoid impacts on wildlife.

The various measures used to reduce impacts of development on wildlife are explored in Chapters 5 to 11. The types of measures included in Chapters 5 to 11 have been grouped according to their function as outlined in Table 1.1. The measures provided in Chapters 5 to 10 are intended to better integrate a development into the natural environment so it is more sensitive to wildlife. Measures that may assist *ex-situ* conservation of wildlife are explored in Chapter 11.

Table 1.2.   Information included in the quick reference tables

What:	A brief description of the measure and a short explanation of how it is intended to reduce an impact of development on wildlife.
Where:	The circumstances under which the measure is used to reduce an impact on wildlife. Examples of the typical types of developments that could incorporate the measure are provided (e.g. mining, roads or residential developments).
Pros:	Possible advantages of using the measure in terms of reducing an impact of development on wildlife.
Cons:	Possible disadvantages of using the measure in terms of reducing an impact of development on wildlife.
Cost:	The cost of applying the measure to a development is indicated as low, medium or high relative to other measures in this book. Broad categories are provided because costs can vary substantially depending on the particularities of the application including the characteristics of the receiving environment.
Companion measures:	Other measures described in this book that are sometimes suitable for use in conjunction with the measure to further reduce an impact of development on wildlife.
Alternative measures:	Alternative measures that could be applied to fulfil a similar function. The alternative measures may be less or more effective for a particular application.

Individual profiles are provided for a range of measures. For each measure, we provide a description and give an indication of its potential application. We explain its intended purpose and reasons why the particular measure is worth considering, among other things. The current usage of each measure in Australia and overseas is also discussed.

We provide a quick reference table for each measure to make it easier for the reader to choose an appropriate solution for their predicament. The quick reference tables outline a range of aspects that are considered when selecting appropriate measures. An explanation of each aspect considered in the quick reference tables is provided in Table 1.2.

A measure is not worth applying unless it is likely to be effective. Evaluating the effectiveness of a certain measure depends on its intended objective. For example, the use of a given measure might be sufficient to reduce an impact on a few individuals but it may not be enough to ensure that a local population persists. Hence, objectives for measures need to be realistic and achievable. We scrutinised the effectiveness of each measure by delving into relevant scientific literature whenever it was available. Sometimes only anecdotal evidence exists and we acknowledge that more research, monitoring and evaluation may be required to establish the effectiveness of some measures. We also provide examples of factors that can influence the effectiveness of the measures, such as its design, placement and maintenance.

When choosing a measure, it is important to take into account whether it could have any adverse effects on the environment. How will the measure affect the surrounding ecosystem? Will it favour particular species and disadvantage others? Possible adverse effects on the environment are discussed for each of the measures included in this book.

After exhausting options to avoid or minimise impacts, there is usually a residual impact on wildlife. This is often associated with loss of habitat. In recent years, it has been common practice for developers to compensate for this factor by providing an environmental offset. This is discussed further in Chapter 12.

Once a measure has been chosen, its performance can be monitored to determine whether it has successfully reduced the impact of the development on wildlife. The use of monitoring programs as an evaluation tool is discussed in Chapter 13. If monitoring and evaluation reveals that the measure has not been successful, it may be varied or replaced with an alternative one. This approach is based on the process of 'adaptive management' as described in Chapter 14.

2

Potential impacts of development on flora and fauna

Types of impacts

Developments can have an impact on terrestrial flora and fauna in a variety of ways. The biological level (e.g. ecosystem, population, species and genetics) impacted, as well as the nature and magnitude of an impact, varies with the type of development, its design and the receiving environment. Some examples of different types of developments are listed in Table 2.1, some of which are also shown in Figure 2.1.

Table 2.2 contains some examples of the types of impacts that can result from development and are relevant to the measures discussed later in this book. Developments can impact wildlife directly (e.g. vehicle collision resulting in the death of an animal) or indirectly (e.g. weed invasion following vegetation clearance), and can occur singularly but more often as a combination of interrelated impacts. These impacts can reduce the numbers of individuals within a population, so the population becomes non-viable, or, worse, threaten a species with extinction.

Climate change can exacerbate many of the potential impacts listed in Table 2.2. Fragmentation of habitat, for example, can put a fauna species at risk due to restrictions on movement and habitat use. However, some species are likely to be at even greater risk in a changed climate because, if individuals are unable to move through fragmented landscapes, they will be unable to reach suitable habitat as their current habitat becomes unsuitable due to changes in climate.

Loss of native vegetation

Loss of native vegetation due to land clearance is a key impact on native wildlife in Australia and can influence many of the other impacts listed in Table 2.2. Loss, fragmentation and degradation of habitat for wildlife; interruption of ecosystem processes; edge effects; changes in

Table 2.1. Example of different types of development

Development	Example
Residential, commercial	Houses, buildings
Infrastructure	Roads, rail, powerlines, pipelines, dams and irrigation
Mining	Open cut, underground
Agriculture	Grazing, cropping, orchards
Energy	Wind farms, solar farms
Tourism	Parks

Figure 2.1. Example of types of developments: (a) powerlines, (b) open-cut mining, (c) intense agriculture, (d) urban development, (e) water pipeline, (f) livestock grazing, (g) roads and (h) wind turbine (Photos: J Gleeson)

Table 2.2.   Examples of impacts on wildlife that can result from development

Impacts	Explanation
Loss of native vegetation	Loss of native plant species in an area.
Physical trauma to animals during land clearing	Land clearing resulting in injury or fatality of animals.
Loss of habitat	Loss of habitat (i.e. environmental factors that permit an organism to survive and/or reproduce) for a particular species.
Habitat fragmentation	Subdivision of habitats into smaller segments.
Habitat degradation	The reduction of quality of an area of habitat for a particular species.
Interruption of ecosystem processes	Interruption of processes that sustain ecosystems (Chapter 3) leading to a loss of functioning and health of an ecosystem.
Edge effects	The change in biological and physical conditions at ecosystem boundaries.
Changes in animal behaviour	Changes in the routine behaviour of animals (e.g. nesting or foraging).
Environmental weeds	Introduced plants that pose a threat to wildlife.
Exotic animals	Animals introduced since Europeans settled Australia, many of which pose a threat to native wildlife.
Artificial lighting	Night lighting at development sites can disrupt foraging behaviour, reproduction and movement of particular animals.
Collision threat	Structures (e.g. buildings) or transport vehicles (e.g. cars or planes) can present a collision threat to animals.
Miscellaneous hazards	Hazards presented by particular developments that can result in wildlife injury or fatalities.

animal behaviour and proliferation of environmental weeds or exotic animals can all be linked to the loss of native vegetation. Most types of development have the potential to result in loss of native vegetation, depending on where they are constructed.

Physical trauma to animals during land clearing

Clearing land in the direct footprint of a development can result in physical trauma (e.g. injuries or fatalities) to animals using the habitat at the time. Most types of development that propose to clear habitats have the potential to result in fauna mortality during land clearance. This is discussed further in Chapter 6.

Loss of habitat

Loss of habitat for wildlife can happen when vegetation is removed, but also through a number of other ways, depending on what constitutes habitat for a particular species. For example, destruction of roost caves represents a loss of habitat for cave-dwelling bat species. When habitat is lost, displaced fauna may not survive due to shortages of available habitat resources (e.g. forage, roosting or breeding resources) in the remaining environment. Similar to native vegetation loss, most types of development can potentially result in loss of habitat, depending on where they are situated.

Habitat fragmentation

Habitat fragmentation, or 'subdivision of habitat', is a common impact arising from development and can result in barriers to movement and dispersal (Lindenmayer and Fischer 2006;

van der Ree *et al.* 2008a). Habitat clearance and physical impediments are the most common causes of habitat fragmentation. Habitat fragmentation can cause division of wildlife populations, isolation of key habitat resources, loss of genetic interchange and ultimately the extinction of species as it reduces the viability of populations (e.g. Hobbs and Yates 2003; Watson *et al.* 2003). The degree to which a development fragments or isolates habitat varies depending on the particular faunal groups or species concerned. Impacts from habitat fragmentation can vary temporally as some species move in relation to a particular life cycle stage (e.g. species may move into an area for breeding) and/or seasonal changes to habitats (e.g. rainfall filling an ephemeral wetland may initiate species moving to the wetland). Fragmentation due to roads and tracks also allows greater access by feral predators that use them (e.g. foxes).

Habitat degradation

Habitat degradation can ultimately lead to habitat loss. Degradation of habitat can be caused by interruption of ecosystem processes, edge effects, environmental weeds, exotic animals and artificial lighting. Processes that simplify habitats can result in habitat degradation (e.g. removal of trees below a powerline). Physical changes to the abiotic environment (e.g. light, noise and soil compaction) can also degrade habitats.

Interruption of ecosystem processes and natural disturbance regimes

Ecosystem processes sustain functioning ecosystems. Such processes include climatic processes, primary productivity, hydrological processes, nutrient cycling, interactions between species (e.g. competition and predation), movements of organisms and natural disturbance regimes (e.g. fires and floods) (Bennett *et al.* 2009; Auld and Keith 2009). Ecosystem processes are often complex and potential impacts on them are difficult to quantify. Most developments in natural settings have the potential to interrupt ecosystem processes. This is explained further in Chapter 3.

A potential consequence of development is an increase or decrease in fire frequency. It is well established that fire is an important component of many Australian ecosystems. Particular fire frequencies and intensities are needed to maintain natural vegetation communities (also called plant communities) (assemblages of native plants in a relatively uniform patch) and species in the landscape (Tolhurst *et al.* 1992; Driscoll *et al.* 2010). Inappropriate fire frequency and intensity can lead to certain species being lost from an area. Some flora species require particular fire regimes to stimulate seed germination (e.g. some sclerophyll plants), while others are destroyed when burnt (e.g. some rainforest plants).

Edge effects

In developed landscapes, the edges of many types of habitat (e.g. forest and woodland) tend to be subject to increased levels of light, noise and dust compared with the middle of the patch. This alters the characteristics of the habitat at the edges in such a way that may ultimately result in a change in species composition. 'Edge effect' is a term used to describe the changes in biological and physical conditions that occur at ecosystem boundaries. Although most edge effects reduce with distance from the edge, the distance is highly variable and under-researched in Australia.

Edge effects are generally greater on remnant patches of vegetation with a low area to edge ratio (i.e. a patch of vegetation with a relatively large perimeter). Most developments adjoining wildlife habitat have the potential to create or add to edge effects. Land clearance can also increase edge effects, particularly when the area to edge ratio of habitat is reduced.

Changes in animal behaviour

Behavioural changes in animals can occur in response to the physical presence of a development or due to interaction with people at a development. There are different types of behavioural changes possible, including changes in choice of foraging location and mating behaviour, which may ultimately lead to changes in species composition and even the functioning of ecosystems. Many types of development can result in behavioural changes in animals.

Environmental weeds

Environmental weeds are plant species from a different locality that invade natural areas and adversely affect the indigenous flora and fauna. It is generally well accepted that weeds are a considerable threat to nature conservation as well as an economic problem worldwide. Depending on the species, they can increase shading, compete with native plants for nutrients, smother native plants (e.g. vines) or chemically suppresses their germination and/or growth (allelopathy). Many developments have the potential to escalate environmental weed progression, including those with gardens that harbour weeds, especially when adjoining habitat areas. However, it should be remembered that developments do not need to be directly adjoining natural areas to cause an impact because birds and other vectors disperse many varieties of weeds.

Exotic animals

Exotic animals are vertebrate and invertebrate fauna that have been introduced to Australia since European settlement. Examples of exotic animals in Australia include rabbits (*Oryctolagus cuniculus*), cane toads (*Bufo marinus*) and fire ants (*Solenopsis invicta*). Exotic animals have degraded many natural habitats. They compete with native species for resources and some prey on native fauna. Many types of developments have the potential to result in local proliferation of exotic animals.

Artificial lighting

Artificial lighting can cause disruption of foraging behaviour (Stone *et al.* 2009), increased potential for collision with made structures (Ogden 1996; Longcore and Rich 2004; Drewitt and Langston 2008) and disruption of reproduction and movement (e.g. Salmon 2003, 2006; Rodriguez and Rodriguez 2009). The effect of artificial lighting on most Australian native fauna has not been sufficiently studied. Nonetheless, inferences can be made regarding the susceptibility of each species by understanding the species' biology and ecology in relation to how they use the nocturnal environment. Australian animals reported to have been impacted to some degree by artificial lighting include marine turtles (Limpus 2008) and sugar gliders (*Petaurus breviceps*) (Shannon 2007). Artificial lighting is used in almost all developments, usually to allow the development to operate at night or for security.

Collision threat

Animal–vehicle collision (or 'vehicle strike') is perhaps the most obvious example of fauna mortality resulting from direct human interaction. Fatality of fauna caused by collisions with motor vehicles is a direct effect that reduces animal populations (Coffin 2007). In Australia, many thousands of collisions between vehicles and animals occur each year (Rowden *et al.* 2008; Ramp and Roger 2008). In 1985, for example, the estimated annual mortality rate for Australian reptiles and frogs was 5 480 000 (Ehmann and Cogger 1985). Animal–vehicle collision is a major source of frog mortality and is thought to have contributed to their global decline (Glista *et al.* 2008). In the United States of America, a staggering one million vertebrate animals are estimated to be killed per day on roads (Lalo 1987). Road-kill is a term used to describe the carcass of an animal resulting from animal–vehicle collision.

Birds are particularly susceptible to collision with structures at development sites (bird strike). Birds have been documented to collide with powerlines (McNeil *et al.* 1985; Janss 2000; Rubolini *et al.* 2005), wind turbines (Smales *et al.* 2005; Smales 2006; Hull and Muir 2010) and buildings (Ogden 1996; Klem 2009).

Miscellaneous hazards

Development sites can present other hazards to fauna that result in injury or fatality. Such hazards include uncovered holes that animals could fall into (e.g. wells or mine shafts) and storages with hazardous materials (e.g. tailings storage facilities at a gold mine).

Assessing impacts

Many government and industry policies are directed at developments to conserve biodiversity (i.e. all aspects of biological diversity, including at the genetic, species and ecosystem levels). Undertaking development and achieving successful conservation of biodiversity relies on correctly evaluating potential impacts and implementing measures to tackle them. To diagnose the full range of impacts, possible 'impact pathways' (i.e. the means by which impacts can occur) must be considered. For example, when assessing the impact of environmental weeds at a development site, it is necessary to consider how the weeds can spread throughout the site. In this case, the impact pathway might be through off-road vehicles driving in and out of weed-infested areas.

Adverse impacts on wildlife can occur simply because assessors fail to identify that a particular environmental feature needs to be avoided or do not identify relevant impact pathways. This is an ongoing challenge because potential impacts on species are not always readily identifiable. Impact pathways can vary temporally (e.g. an action now may materialise as an impact in the future) and spatially (e.g. an action undertaken at one location may cause an impact at another location). Measures designed to reduce impacts at the planning stage for a new development may need to evolve over time.

It is important to consider multiple spatial scales, especially when investigating the impacts from habitat loss and modification. Relevant ecological processes should be identified as well as how they influence species (Bennett *et al.* 2009; McDonald and Williams 2009). Although this can be challenging in highly modified landscapes, it is also more important to do so. This is because the viability of the remnant habitat in such an area can be reliant on the remaining ecosystem processes that sustain it. If there has already been a significant decline in ecosystem health, any additional change can exacerbate that decline, perhaps past an ecological threshold: a point at which small environmental changes produce substantial responses in ecosystem state or function (Groffman *et al.* 2006; Samhouri *et al.* 2011).

Cumulative ecological impacts should be assessed by identifying any environmental stressors already operating in a particular area and considering how those stressors, in conjunction with a new development, will affect the environment. Data sharing between developers is very useful to identify existing environmental stresses. Unfortunately, this does not often happen due to the competitive nature of most businesses. Ideally, government regulatory planning in Australia should aim to close this gap.

Assessing potential impacts of development is an iterative process. Over time, our understanding of how existing types of developments impact on wildlife continues to improve and new types of developments are created posing new types of impacts. For example, the increasing use of wind farms in Australia during the last 10 years has sparked studies of the risk of collision posed by wind farms to birds and bats (e.g. Smales *et al.* 2005; Smales 2006; Hull and Muir 2010) (Figure 2.1h). Similarly, the preferred measures to reduce impacts on wildlife are also likely to change as new information comes to light.

Ecological impact assessment involves identifying and evaluating potential impacts on biodiversity, managing these impacts and assessing the success of measures used to reduce the impact of development on wildlife (Environmental Institute of Australia and New Zealand Ecology 2010). Prior to a development going ahead in Australia, ecological impact assessments

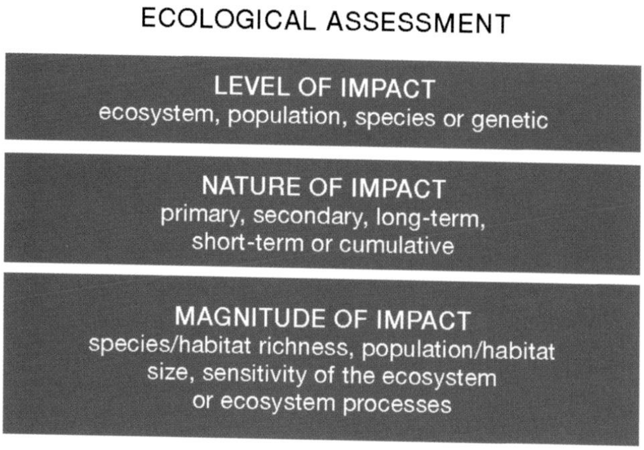

ECOLOGICAL ASSESSMENT

LEVEL OF IMPACT
ecosystem, population, species or genetic

NATURE OF IMPACT
primary, secondary, long-term,
short-term or cumulative

MAGNITUDE OF IMPACT
species/habitat richness, population/habitat
size, sensitivity of the ecosystem
or ecosystem processes

Figure 2.2. Components of ecological impact assessment (Diagram: J Toich)

are often required by law to identify the potential level, nature and magnitude of impacts (e.g. Commonwealth of Australia 2007a) (Figure 2.2) and any need for impact avoidance, mitigation and offset measures.

An environmental risk analysis process has a role to play in the assessment of ecological impacts. This process involves identifying: firstly any impacts that may potentially occur; secondly the likelihood of the potential impact occurring; thirdly the potential magnitude of the impact; and lastly suitable management options (Elliot and Thomas 2009).

3

Selecting measures to reduce impacts of development on wildlife – why an understanding of ecology is fundamental

An understanding of ecology – that is, the study of the relationships between organisms, groups of organisms and their biotic (living) and abiotic (physical and chemical) environments – is fundamental for:

- identifying whether a measure is required to reduce an impact of development on wildlife
- selecting an appropriate measure to avoid or minimise the identified impact
- ensuring that the measure is implemented as intended.

Looking at each situation from an ecological perspective gives us an insight into how wildlife is likely to be affected by a development. This is a core component of conservation biology, which involves the application of ecological principles to tackle conservation challenges (Perlman and Milder 2005).

Measures that are selected and applied without an understanding of ecology risk failing, and that is one reason why this book does not replace the need for a qualified and experienced ecologist to evaluate each individual development situation. In some situations, incorrect application of a measure can also be damaging to the environment. For example, the haphazard installation of fauna exclusion fencing along roads can create barriers that isolate individuals of a population, perhaps leading to over-abundance of fauna in a restricted habitat and potential inbreeding (Jaeger and Fahrig 2004; Balkenhol and Waits 2009).

Throughout this book we assume that the reader has a basic understanding of ecology. However, we have provided a brief overview of some key ecological concepts in this chapter for those who may benefit from a brief introduction or refresher. An in-depth review of the concepts central to ecology is beyond the scope of this book. There are several excellent texts that provide a background on ecology, including *Ecology: From Individuals to Ecosystems* by Begon *et al.* (2005) and *Ecology: An Australian Perspective* by Attiwill and Wilson (2006). Lindenmayer and Burgman's (2005) *Practical Conservation Biology* is also well worth reading.

Ecosystem processes, function and resilience

Management of wildlife requires an understanding of the broader ecosystems and processes that sustain them. An ecosystem is a biotic community (comprising of plants, animals and micro-organisms such as fungi and bacteria) and the physical environment in which it occurs.

The processes that sustain functioning ecosystems are both spatially and temporally dynamic. A challenge for environmental professionals is to assess the impacts of development on ecosystem processes while keeping in mind that these impacts can take years, or even lifetimes, to eventuate. An example of a natural vegetation community, and one of the ecosystem processes that sustains it, is shown in Box 3.1.

Box 3.1.   An ecological process that sustains plant communities in linear groves

A development that interrupts ecological processes can affect the functioning of an ecological community. A good example of this is how linear groves of vegetation dependant on sheet wash can be affected by the physical placement of developments that obstruct surface water flow. Sheet wash is an ecological process that occurs where the gradient of the land is so slight that rainwater runoff occurs as a sheet of water over the land surface. In many semi-arid and arid ecosystems across Australia, a number of different plant communities occur on these types of substrates, some as linear groves of vegetation (White 1971; Tongway and Ludwig 1990).

The linear groves of vegetation are arranged perpendicular to the contour of the land and the direction of surface water flow. Between the groves, there is often bare ground or grasslands (White 1971; Anderson and Hodgkinson 1997). The vegetation groves capture nutrients and loose vegetative material carried in the water. Captured vegetative material breaks down and improves the soil within patches, driving biological productivity (Cook and Dawes-Gromadzki 2005). Select mulga (*Acacia aneura*)-dominant woodlands occur on sheet wash surfaces and, in this form, are broadly referred to as banded mulga (White 1971; Anderson and Hodgkinson 1997) (Figure 3.1).

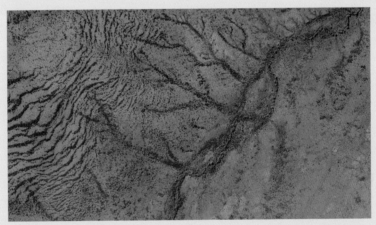

Figure 3.1.   Aerial photograph of banded mulga (*Acacia aneura*)-dominant woodlands, Western Australia (Photo: G Gorman)

Because the grove formation is dependent on the sheet flow process, developments that obstruct the flow of sheet flow (e.g. roads, rail and pipelines) can affect the ecological community downslope of the development. Minimising the spatial footprint or re-positioning the development may avoid this impact.

In built environments, anthropogenic processes can sustain species in artificial ecosystems. For example, a threatened aquatic moss *Fissidens berteroi* occurs in small drainage pipes near Onehunga, New Zealand and is maintained at the site by controlled water release, substituting natural hydrological processes (New Zealand Plant Conservation Network 2011). Similarly, the green and golden bell frog (*Litoria aurea*) habitat surrounding artificial frog ponds at Sydney Olympic Park is subject to an intense landscape management program to maintain habitat values (Darcovich and O'Meara 2008) (Chapter 10). Maintenance of artificial ecosystems can be costly and laborious, making it less desirable than conserving the species in a natural, self-sustaining ecosystem.

It is important to identify the functional roles that species play in maintaining ecological processes because loss of those species can result in loss of or significant changes to key components of an ecological community. For example, *Pararistolochia praevenosa* is a near threatened native vine in rainforests in South-East Queensland. Loss of this vine would result in the loss of the vulnerable Richmond Birdwing Butterfly (*Ornithoptera richmondia*) because its larva depends on the vine as a source of food (Sands 2008).

Natural disturbance regimes (e.g. flooding and fire) are often an integral part of ecosystem function (Perlman and Milder 2005). It is important that measures implemented to reduce impacts of development on a particular species do not unintendedly impair natural disturbance regimes. Similarly, care should be taken to ensure that natural disturbance regimes do not undermine the functioning of any measures implemented. For example, a fauna exclusion fence designed to protect frogs from danger could be rendered ineffective by natural flooding regimes if installed in a flood-prone area. The potential for high impact natural disturbances that have a low likelihood of occurring should also be considered and contingencies should be put in place to make sure that impacts on wildlife remain minimal after the natural disturbance.

Building ecosystem resilience is a priority recognised in *Australia's Biodiversity Conservation Strategy 2010–2030* because natural systems must be resilient to withstand environmental stresses (National Biodiversity Strategy Review Task Group 2010). Ecosystem resilience is the ability of an ecosystem to adapt to changes and disturbances while maintaining the same basic function, structure and interactions. For example, a 100-year-old cropped paddock will have very low resilience if crops are removed because there will be poor storages of native seeds in the seed bank. On the other hand, a native pasture grazed intermittently would have moderately higher resilience because the seed bank is likely to contain a lot more native seeds. Various measures covered in this book can be used to build ecosystem resilience, such as retaining or creating natural habitat linkages (Chapter 8) and revegetating and restoring ecosystems (Chapter 10).

Species present and absent, and their habitat

Identifying species that could be present in and near a development site and understanding their habitat requirements is fundamental when choosing appropriate measures to reduce impacts. Species can require different habitat resources at different stages of their life cycle, so measures may need to be directed towards a particular life cycle stage. With this type of information, it is possible to apply measures that are targeted to tackle an impact on a particular species, such as a threatened or keystone species (i.e. species-specific measures). In some situations, a measure used to conserve a particular species can disadvantage another species present at the site. For example, a fauna exclusion fence might be beneficial to protect particular fauna species, but erecting a fence will be disastrous if the vegetation clearance required to construct the fence causes the loss of a population of threatened plants.

It can be equally important to identify which species are not present in an area of potential habitat and investigate reasons why they are not present. This information is useful not only in assessing impacts from a development but also to identify what measures could be applied to improve the prospects of the habitat that will remain.

The ecological role that a particular species has in the food chain and the ecological relationships between species are also considerations when determining management options. Although a development may pose a direct threat to a particular species in the food chain (e.g. herbivore), other species at higher trophic levels (e.g. carnivore) could be indirectly affected, requiring multiple management measures.

The effectiveness of many of the measures described in this book depends on the type and complexity of the habitat in which they are proposed. For example, installing a fauna exclusion fence along an extensive highway in the arid zone of central Australia to prevent vehicle strike with kangaroos would be impractical, and possibly detrimental, because the species' habitat is vast and the impact from constructing the fence would probably outweigh the benefits.

Population ecology

Many of the measures described in this book are implemented with the aim of maintaining viable populations of flora and fauna. Conserving populations works towards conserving species and genetic variation (Lindenmayer and Burgman 2005). Understanding the structure and geographical distribution of a species' population in relation to a proposed development enables more definitive conclusions to be drawn about the likely effectiveness of any proposed measures.

A population is a group of individuals from the same species, commonly forming a breeding unit in which the exchange of genetic material is more or less unrestricted. Other levels of social structuring may also be relevant to population structure (e.g. white-browed babblers *Pomatostomus superciliosus* form family groups; Cale 2003). In some instances, individuals of a species may be present in a population that is part of a bigger metapopulation (a set of populations). These populations are usually somewhat isolated from each other in discrete patches of habitat, but some individuals do disperse between the local populations. Fragmentation can exasperate the occurrence of sub-populations. In extreme cases, if there is no dispersal of individuals between sub-populations, the risk of extinction increases because the small size of each sub-population can make it more vulnerable to natural disturbances such as bushfire, as well as to impacts from development. Populations must be large enough that genetic variation is maintained because a loss in genetic variation can influence the dynamics and persistence of populations.

The relative abundance of a species near or within a development site can influence the measures needed to reduce impacts on that species. If a species is in low abundance, different measures may be more suitable than if the species is in high abundance. In some cases, low abundance may mean that more is required to maintain the presence of the species because it has already suffered decline in abundance, while, in other cases, the species may be naturally in low abundance. The change in composition of populations over time (i.e. population dynamics), is often used in monitoring the success of measures.

Animal behaviour

Understanding the way that an animal reacts to a development is important when choosing measures to reduce impacts. One common behavioural trait that is relevant to ecological impact assessment is an animal's reflex, which is its automatic response to a stimulus. For example, some fauna species have a behavioural avoidance of roads and are unlikely to be at risk of collision with motor vehicles (Jaeger *et al.* 2005; Shepard *et al.* 2008). This may negate the need to provide measures to reduce road collision for the particular species.

Figure 3.2.   Radio-tracking device fitted to a central bearded dragon (*Pogona vitticeps*), Queensland (Photo: M Gleeson)

Habituation occurs when an animal begins to ignore stimulus or otherwise becomes accustomed to stimuli through prolonged and regular exposure. Some individual animals can become remarkably resistant to disturbance caused by development. Although it is generally positive for animals to become accustomed to nearby development, it is an issue when needing to deter fauna from a development site for their safety because the animal may over time ignore the chosen measure (e.g. visual deterrent).

Understanding animal behaviour is particularly important when rescuing fauna during land clearance (Chapter 6), deterring fauna from a development site (Chapter 7) and designing a measure to promote safe movement of fauna (Chapter 8).

Movement and dispersal

The movement of animals and dispersal of plants occurs at several spatial and temporal scales and is critical for the survival and successful reproduction of individuals and persistence of populations. It is necessary to understand how, where and why a species moves in order to apply measures to reduce impact with optimal outcomes. Often this can be inferred by an experienced ecologist. In other cases, radio-tracking (Figure 3.2) (e.g. Ball and Goldingay 2008), genetic techniques (e.g. Simmons *et al.* 2010) and capture-mark-re-capture methods (e.g. Hamer *et al.* 2008) are used to observe the movement of animals. Other techniques used to gain an insight into animal movement include analysing statistics (e.g. road-kill data) to determine movement paths (e.g. Ramp *et al.* 2005a), radar (e.g. Australian Transport Safety Bureau 2008) and modelling (e.g. Shaw *et al.* 2010; Rollan *et al.* 2010).

Some animals have designated movement paths (e.g. many bat species have flyways or flight paths). Often these paths are linked to key habitat areas, such as water, foraging or

4

Avoiding impacts on flora and fauna

It is best practice to first exhaust options to avoid impacts on wildlife before attempting to mitigate impacts or resorting to compensatory measures. By applying measures to avoid impacts on wildlife first and foremost, we can be more confident that a particular species or ecological community will be protected if a development proceeds.

This chapter discusses how effective land use planning for conservation outcomes is an overarching measure for avoiding impacts on wildlife. Conservation planning by defining land uses provides clear boundaries for developers. Conservation planning outside of the reserve network (e.g. national parks and nature reserves) can be as important, if not more important, than conservation in reserves.

At a development site, there is often the opportunity to avoid impacts without compromising the function of the development. Missed opportunity may be due to not identifying potential impacts and the measures to avoid them. Another limitation is gaps in knowledge leading to uncertainty over whether an impact will occur or not. In this situation, the precautionary principle should be applied, as discussed in this chapter.

Incorporating measures to avoid impacts on wildlife at a development site can be more cost-effective than managing impacts, because they negate the need for more laborious mitigation measures and ongoing monitoring. The need to avoid impacts of development on wildlife occurs not only at the initial development planning stage, but throughout the various stages of the development (e.g. design, construction and operation), for the life of the development. On-site measures to avoid impacts on wildlife and to establish wildlife buffers are discussed in this chapter.

Conservation planning

Strategic land use planning at several spatial scales plays an key role in avoiding impacts on wildlife (McIvor and McIntyre 2002; Huber *et al.* 2010), particularly cumulative impacts. Land use planning allows better integration and management of wildlife conservation with human development (Perlman and Milder 2005). It also allows for strategic separation and buffering of conflicting land uses. For example, effective land use planning around airports reduces the likelihood of wildlife–aircraft collisions occurring (Blackwell *et al.* 2009).

As mentioned above, conservation planning outside of the reserve network can be as important, if not more important, than conservation in reserves. The retention and protection of native habitats is a high priority for the conservation of biodiversity in Australia (National

Biodiversity Strategy Review Task Group 2010). Currently, only 11.88% of the Australian continent is protected under the national reserve system (Department of Sustainability, Environment, Water, Population and Communities 2011c), in addition to private off-reserve conservation (e.g. by groups such Bush Heritage Australia and Australian Wildlife Conservancy). This leaves a vast area of Australia subject to other land uses.

Conservation planning is not only used to identify blocks of habitat that are best protected, but also to identify important natural habitat linkages and ecological gradients between habitats (e.g. connectivity across terrestrial areas to freshwater and marine habitats; Beger *et al.* 2010). This connectivity is often needed for the survival and successful reproduction of individuals and persistence of populations.

Managing and Conserving Grassy Woodlands by McIntyre *et al.* (2002) provides a good overview of the principles for conservation planning at the agricultural property scale, with an emphasis on eucalypt woodlands. Conservation planning in agricultural landscapes forms the backbone of sustainable agriculture.

Ecologically sustainable development and the precautionary principle

Ecologically sustainable development is a policy concept integrated into planning policies at various levels of government in Australia. Australia's National Strategy for Ecologically Sustainable Development (Ecologically Sustainable Development Steering Committee 1992) defines ecologically sustainable development as: 'using, conserving and enhancing the community's resources so that ecological processes, on which life depends, are maintained and the total quality of life, now and in the future, can be increased'. The concept involves the consideration of ecological, social and economical aspects in the decision-making process of using natural resources. Ecologically sustainable development is underpinned by the principle of inter-generational equity, which requires natural resources (and wildlife) to be maintained and enhanced for future generations.

Gaps in technical knowledge can obstruct sound decision making in relation to ecologically sustainable development. Where there is a lack of scientific certainty, precautionary measures should be taken to avoid impacts in accordance with the precautionary principle, defined in the Commonwealth *Environmental Protection and Biodiversity Conservation Act, 1999*. The precautionary principle is that 'the lack of full scientific certainty should not be used as a reason for postponing a measure to prevent degradation of the environment where there are threats of serious or irreversible environmental damage'. The following questions are worthwhile asking, particularly in the early stages of planning:

- Is the requirement for the development justified?
- Do the environmental impacts outweigh the gains of the development?
- Are there better alternatives for the development?

In some cases, if impacts on wildlife cannot be sufficiently reduced, it may not be appropriate to proceed with a development despite using mitigation measures (Box 4.1). It will depend on whether the residual impacts that cannot be mitigated or offset are acceptable.

Measures to avoid impacts on wildlife

Sometimes it can be possible to avoid impacts on wildlife in a specific area. Depending on the situation, impacts could be avoided by relocating or re-designing the development, timing the

Box 4.1.   Proposed Traveston Crossing Dam, Queensland

In 2006, potential water shortages led the Queensland Government to propose a new dam, to be called the Traveston Crossing Dam, to be constructed on the Mary River in South-East Queensland. An environmental impact assessment was conducted that identified potential impacts on threatened species of national significance: the Australian Lungfish (*Neoceratodus forsteri*), Mary River Cod (*Maccullochella peelii mariensis*) and Mary River Turtle (*Elusor macrurus*) (Sinclair Knight Merz 2007; Walker 2008; Bunn 2008).

The proposed Traveston Crossing Dam was refused Commonwealth environmental approval due to its proposed impacts (Department of Sustainability, Environment, Water, Population and Communities 2011a). In terms of ecologically sustainable development, in this case, the environmental impact was deemed to outweigh the social and economical need for the dam.

development to avoid impacts, or using or shunning specific practices during construction. Day-to-day management of the development also provides opportunities to avoid impacts on wildlife. These measures are discussed below.

The measures described in this chapter are not always applied to prevent all impacts from occurring. In some situations, the measures will lower the degree of impact on wildlife, but an impact will still occur. For example, selecting a different location for a development may reduce the area of native vegetation needing to be cleared, but some clearance of the vegetation may still be required for the development to proceed.

Many measures to avoid impacts on wildlife relate to the retention of vegetation, topsoil and wildlife habitat. There are many reasons to retain vegetation on a development site, such as improved stormwater management, shade and reduction of heat, wildlife habitat and/or aesthetics (Australian Standard 4970-2009). In addition, retaining native vegetation reduces clearing costs as well as remediation costs for temporary developments (Brennan *et al.* 2005).

Alternative location for the development

Forward planning when selecting a location for a development can be the most important factor in avoiding impacts on wildlife. This relates to the broad concept of conservation planning described earlier in this chapter, but is more specific to the placement of a development within the land use planning framework. Areas of wildlife habitat (e.g. movement corridors or watercourses or areas of high conservation value) can be avoided by locating the development a distance from them (refer to wildlife buffers described later in this chapter). For example, relocating a road away from frog-breeding habitat (e.g. wetlands) may avoid impacts on frogs because the highest rates of frog road mortality occur when a road is located adjacent to breeding habitat (Glista *et al.* 2008; Andrews *et al.* 2008). The alternative locations for the road may have varying impacts on other wildlife, particularly where habitat clearance will result.

When considering alternative locations for a development, the potential impacts should be evaluated at each proposed site. There are many different environmental factors that would be relevant when choosing a site where development would have less impact on wildlife, and these factors would vary with the type of development and the receiving environment. Impacts on wildlife may be *reduced* (rather than *avoided*), depending on the characteristics of the different locations. Impacts on wildlife are generally lower in a location that will:

- retain areas of high conservation value (e.g. threatened and rare species and ecological communities)
- maximise the area of native vegetation and habitat that will remain following the development
- minimise physical incursion on important habitat areas (e.g. core habitat, movement corridors and watercourses)
- maximise the area within patches of remaining vegetation relative to the perimeter of the patch (i.e. the area to edge ratio) to maintain ecological integrity.

Lowe (2009) argues that to become a sustainable society in Australia, there should be no further land clearing of natural areas. He makes a valid point, but this would be very difficult to achieve in practice for all developments because many have restrictions on where they can be sited. For example, mining developments must be located where the mineral resources are located beneath the ground. Other developments could be better focused on infilling and re-using areas that have already been cleared, as Lowe (2009) also suggests. In this case, consideration should be given to whether the cleared land provides habitat services (e.g. movement corridor). Cleared land comprising of grassland located within a mosaic of woodland or forest can actually be beneficial, leading to increased biodiversity in an area.

Alternative designs for development

The design of a development can be equally as important as its location when attempting to avoid impacts on wildlife. The evaluation of alternative designs generally takes place at the outset of a proposed new development. However, for some developments, the requirement for an alternative design may also be identified through the process of adaptive management if monitoring indicates that an existing design is more detrimental to wildlife than predicted. Good design will include measures to avoid impacts on wildlife such as those provided in Table 4.1.

Timing of development

Timing is important ecologically, influencing key life cycle factors such as breeding, feeding, migration and seed production. In some circumstances, it is possible to time construction or operation activities to avoid, or at least minimise, impacts on particular flora and fauna species. For example, if a hibernating bat colony (e.g. in a tree) is disturbed, the bats will take flight and this can lead to premature exhaustion of fat reserves and, consequently, death (Van Dyck and Strahan 2008). In this case, it is best to undertake the activities that may disturb the bats (e.g. vegetation clearance) in a season outside of the hibernation period. Likewise, it is thought that impacts on some frog species can be minimised if land clearance is undertaken when the frogs are alert (i.e. not during winter when the frogs are in torpor) but not breeding (i.e. outside of the breeding season) (Department of the Environment, Water, Heritage and Arts 2009a).

Sometimes it is necessary to delay certain components of a development until more is known about likely impacts (e.g. through a research or investigative study). Construction of a development could be staged to avoid impacts that would otherwise occur if the entire development was constructed at the one time. Many highly mobile animals are able to relocate of their own accord during staged construction if there are sufficient alternative habitat resources. Similarly, progressive rehabilitation and remediation of disturbance areas following disturbance (e.g. mining) is useful for avoiding permanent loss of habitat for nearby species.

Practices used during construction

Construction is the phase of a development that generally results in the greatest area of land disturbance. In addition to the development footprint, contractor camps and material lay-down

Table 4.1. Some impact avoidance measures and examples of how they may be applied to the design of a development

Aspect	Measure	Example
Spatial footprint	Rationalise components of a development.	Build multiple storeys on buildings rather than spreading one storey buildings over a greater area.
	Increase efficiency of the components of a development.	Select a more direct route for linear infrastructure.
	Group components of a development.	Streamline linear infrastructure, such as roads, rail and power, within the same corridor.
	Re-use existing facilities before creating new ones.	Share infrastructure between developments.
Barriers to faunal movement	Avoid placement of structures that can obstruct movement.	Burying pipelines would enable animals to move over them and sufficiently raising pipelines on concrete blocks would enable small animals to pass under them.
	Avoid land clearance that can obstruct movement.	Avoid land clearing that would fragment a natural habitat linkage (Chapter 8).
	Design structures to avoid wildlife interactions.	Installing powerlines underground in areas where collisions by birds and bats is a high risk.
Hazards to wildlife	Avoid use of the particular components of a development that are hazardous to wildlife.	Avoid use of glass screens along roads.
Interrupting ecosystem processes	Avoid activities that can interrupt ecosystem processes.	Site developments to avoid impacting hydrological processes and nutrient cycling necessary for ecosystem functioning (Box 3.1).
Watercourses	Avoid activities that may directly or indirectly impact watercourses.	Maintain setback distance (i.e. 'buffers') from riparian vegetation and catchment drainage lines.
Artificial lighting	Site illuminated areas away from natural areas susceptible to impacts of artificial lighting.	Maintain setback distance from a beach to ensure that light does not confuse hatchling marine turtles trying to make their way to the water (Chapter 9).
Human–wildlife interaction	Design a development away from natural habitats.	Locate residential estates away from important conservation sites or erect fences between houses and natural areas.
	Design a road to reduce blind spots to reduce road-kill.	Increase visibility with a clearway at the side of the road.
	Re-use existing facilities before creating new ones.	Duplicate existing roads rather than create new roads through natural areas.
Areas of high conservation value	Avoid areas of high conservation value.	Relocate support infrastructure for a mining development to avoid destroying a small group of threatened plants.

areas are often required. Sometimes, it is possible to incorporate the areas required for temporary use during construction into the overall footprint of the development.

Some practices used during construction can increase the risk to fauna. For example, in 2004, a truly astounding 6500 animals, mostly reptiles, reportedly died in a trench dug for a

Table 4.2. Some measures to avoid impact of construction on wildlife and examples of how they may be applied

Measure	Example
Avoid disturbing land outside the final development footprint.	Place areas required for temporary use during construction within the overall footprint of the development (Box 4.2).
Locate and operate construction equipment to minimise soil disturbance and compaction.	Minimising the potential for soil compaction, particularly at or near the root zone, is a recognised impact-avoidance measure specific to the critically endangered spiny rice-flower (*Pimelea spinescens subsp. spinescens*) (Department of the Environment, Water, Heritage and Arts 2009b).
Delineate construction boundaries.	A fence or barrier can be used to avoid incidental disturbances to key areas outside of the construction footprint (Figure 4.1).
Avoid construction methods that are hazardous to wildlife.	Trenchless installation of pipelines by subterranean tunnelling avoids the potential for fauna to be trapped in a trench dug for installing underground pipes.

442 km pipeline development in northern Western Australia (ABC News 2004). In this example, minimising the length of the trench open at any period of time could have prevented such an adverse impact (Australian Pipeline Industry Association 2009).

Figure 4.1. Fencing and signage to prevent disturbance to the root zone of a tree. Australian Standard 4970-2009 provides an overview of protecting trees on development sites (Photo: J Gleeson)

Box 4.2. Construction of the Skyrail Rainforest Cableway – an example of avoiding excess clearance during construction

The Skyrail Rainforest Cableway in Cairns, Queensland, spans 7.5 km across the Barron Gorge National Park, which is registered as a World Heritage rainforest within the Wet Tropics World Heritage Area. The eco-tourism attraction comprises cable-strung gondolas that take visitors across the tree tops, allowing them to view the rainforest from above. It was reported to be the longest cableway in the world when it commenced construction in 1994 (Beeh 1994), with cables strung between 32 towers, up to 40.5 m high (Skyrail Rainforest Cableway 2011).

Due to the obvious sensitivities of constructing a large-scale commercial development within a World Heritage rainforest, there was a debate over the pros and cons of the installing the cableway (Beeh 1994). In order to reduce impacts on wildlife, a range of measures were adopted. Perhaps the measure that lessened impacts the most was using Russian Kamov helicopters to carry equipment and materials to the tower sites (Skyrail Rainforest Cableway 2011) (Figure 4.2). This use of helicopters meant that large tracks of rainforest between towers did not need to be cleared to facilitate construction. Helicopters were also used to string the cables between the towers, a method also more widely used by power companies to string powerlines across sensitive areas.

Figure 4.2. Construction of the Skyrail Rainforest Cableway using a helicopter (Photos: Skyrail Rainforest Cableway)

There are numerous standard construction methods adopted by industry to minimise environmental impacts in general (e.g. stringent erosion control; Witheridge 2010). However, there are also more specific construction methods that can be used to avoid impacts on wildlife. Some measures to avoid impacts of construction on wildlife and examples of how they might be applied are listed in Table 4.2.

Management of operation of the development

Even after a development has been constructed, wildlife can still be adversely impacted during the day-to-day operation of the development. Impacts from poor management can occur progressively with minimal consideration for the long-term outcomes of the management decisions. Relevant measures to prevent impacts on wildlife at the operations stage can usually be grouped into measures for managing native vegetation and those for preventing hazards to wildlife. Some of these measures and examples of how they might be applied are outlined in Table 4.3.

Wildlife buffers

'Wildlife buffers' are areas used to separate areas with native wildlife from incompatible land uses and are a key concept in conservation planning. Wildlife buffers have been used to protect a

---

Box 4.3. A buffer between a mine and a cave used by the orange leaf-nosed bat (*Rhinonicteris aurantia* subsp. Pilbara)

BHP Billiton Iron Ore Pty Ltd's Goldsworthy iron ore mining operations, located in the northern Pilbara region of Western Australia, comprise a series of open cut mines. In 2004, caves suspected to contain significant colonies of the vulnerable orange leaf-nosed bat (*Rhinonicteris aurantia* subsp. Pilbara) were identified by the company near a proposed mine extension area. The finding was significant because limited natural roosts were known in the Pilbara (Armstrong 2001). This bat requires specific roosts that are hot (28–32°C) and humid (96–100% relative humidity) and contain open water or seeps inside the cave (Armstrong 2001; Churchill 2008).

In recognition of the importance of the caves to the orange leaf-nosed bat, a buffer from the mine was established. The objective of the buffer was to avoid significant impacts of the mine on this species. The buffer distance was selected based on local observations of bats persisting at caves near other mining infrastructure (e.g. active haul roads). This distance was then increased conservatively to a buffer of 400 m at the proposed mine extension area. The buffer was established by defining the zone on mine plans, installing signage at the edge of the buffer and by promoting awareness of the buffer through mine inductions for new staff and visitors.

To evaluate the effectiveness of the buffer, BHP Billiton Iron Ore has undertaken ongoing management (e.g. maintaining signage) as well as a standardised quarterly monitoring program based on activity patterns at caves adjacent to mining and nearby control sites. Monitoring has shown that the orange leaf-nosed bat is continuing to use the buffered caves and data suggest that fluctuations in activity amongst surveys are the result of natural behaviour rather than the influence of mining (BHP Billiton Iron Ore and Specialised Zoological unpublished data).

Table 4.3. Examples of impact avoidance measures and examples of how they may be applied to development site management

Aspect of development site management	Measure	Example
Native vegetation management	Retain any native vegetation	A shrub layer can often be maintained within a powerline corridor to help minimise loss of species richness, limit colonisation opportunities for weeds and provide continuity of habitat for various species (Clarke and White 2008).
	Retain a mosaic of native vegetation	Mosaic clearing is used in forestry operations to provide continuity of habitat for various fauna species (e.g. birds).
	Retain connectivity of canopy vegetation	Goosem (2000) and Goosem et al. (2001) have shown that maintaining connection of the tree canopy above linear clearings reduces barrier effects for small mammals in rainforest areas.
	Allow fallen trees, branches and leaf litter to accumulate	These are important components of natural systems and provide nutrients and habitat. It should be noted, however, that fuel-load management is sometimes necessary to reduce the risk of bushfires occurring to protect human amenities (e.g. houses and businesses) (New South Wales Rural Fire Service 2006).
	Restrict human access	Fences can be used to discourage access of people into areas adjoining residential estates.
	Prevent intense grazing	Good grazing management can prevent degradation of habitat areas and suppression of native flora regeneration caused by livestock.
	Retain standing dead trees	Dead trees are used for roosting and nesting by many native birds and bats in Australia. By retaining such trees, impacts on birds and bats that use the dead trees as habitat can be avoided.
	Delineating vegetation clearance boundaries	A fence or barrier can be used to avoid incidental disturbances to key areas outside of the development footprint.
Preventing hazards to wildlife	Identify and remove hazards to wildlife	Holes can act as 'pitfall traps' by trapping ground-dwelling fauna, particularly reptiles (Lloyd et al. 2002; Pedler 2010). Covering or capping holes that present a risk can remove this hazard.
	Minimise use of hazardous substances	Leading mining practice is to minimise the use of cyanide – a hazardous industrial chemical used in the process of gold extraction – by recycling as much cyanide as possible (Commonwealth of Australia 2008).
	Use pesticides and fertilisers as directed and away from watercourses under suitable climatic conditions.	Barton and Davies (1993) found that a buffer greater than 50 m was needed to protect riparian and aquatic areas from pesticides in eucalypt plantations.
	Control domestic pets such as cats and dogs	Banning cat ownership in environmentally sensitive areas (Lilith et al. 2006) and keeping dogs and cats inside at night reduces impacts on nocturnal fauna.

particular feature (e.g. cave habitat; see Box 4.3), to avoid edge effects impacting a particular area or core habitat (Dignan and Bren 2003) and to prevent human presence disturbing routine activities of fauna (Rodgers and Smith 1995; Blumstein *et al.* 2003; Fernandez-Juricic *et al.* 2004).

We use the term wildlife buffer to describe a setback distance from a particular habitat. Others have used the term to describe the minimum width of habitat in which a species exists (e.g. the width of riparian vegetation along a stream to maintain a species in the riparian habitat; Hylander *et al.* 2004). Most wildlife buffers can, and often do, still provide suitable habitat for a range of species.

The effectiveness of a buffer can be improved by managing a development in such a way as to contain impacts as much as possible to the site (e.g. by adding controls on sound and light sources). Optimum buffer distances vary depending on the:

- intended purpose of the wildlife buffer (e.g. to reduce or prevent an impact)
- impact intended to be buffered and its characteristics
- characteristics of the habitat and species to be buffered
- characteristics of the surrounding environment (e.g. topography)
- composition (e.g. floristic and age) and structure of the wildlife buffer.

The success of a wildlife buffer is dependent on setting a clear objective from the outset, careful design and ongoing management, as well as a means to evaluate its effectiveness (Weston *et al.* 2009). Wildlife buffers that are sub-optimal may reduce the impact to a degree, yet still adversely impact the species or its habitat (Cockle and Richardson 2003). Ongoing management of a wildlife buffer is often needed to ensure its continued effectiveness.

The optimum width for a wildlife buffer, or setback distance, is often inferred from studies of movement and habitat use, as well as from studies on edge effects. Some examples of varying buffer widths suggested for a range of impact, buffer and taxa combinations are provided in

Table 4.4.   Examples of varying buffer widths

Buffer width (m)	Example
30	Lemckert and Brassil (2000) concluded that a buffer of 30 m from a stream should be effective in protecting the breeding habitat generally used by the endangered giant barred frog (*Mixophyes iteratus*) from logging in northern New South Wales.
>50	Barton and Davies (1993) recommended a setback distance of greater than 50 m from riparian and aquatic areas to buffer the use of pesticides in eucalypt plantations.
>100	Shirely and Smith (2005) recommended a setback distance of greater than 100 m to maintain forest interior bird species along watercourses within coastal temperate forest to buffer forestry in Vancouver Island, Canada.
100–180	Rodgers and Smith (1995) recommended a setback distance of approximately 100 m for wading bird colonies and 180 m for tern and skimmer colonies to buffer pedestrians and motor boats in Florida, United States of America.
200	Hamer *et al.* (2008) inferred a minimum 200 m buffer was required around clusters of water bodies within close proximity to each other for conservation of the green and golden bell frog, New South Wales.
4000	Jotikapukkana *et al.* (2010) considered a 4 km forest buffer was required around a wildlife sanctuary in Thailand to reduce the impact of humans and domestic animals on the wildlife in the sanctuary.

Table 4.4. Unfortunately, there is a lack of published scientific data on the effectiveness of wildlife buffer distances for most Australian wildlife.

The establishment of a wildlife buffer can be constrained by availability of land tenure and characteristics of the land that would form the wildlife buffer. In some circumstances, it is necessary for the land to be revegetated to create a sufficient buffer. In these cases, the resulting time lag must be taken into account when scheduling the development.

Vegetation buffers are generally not useful for species that use edge habitat (e.g. birds that live on forest/woodland boundaries) because the habitat is effectively brought closer to the development. In these circumstances, a setback distance without dense revegetation would be more applicable.

5

Modifying human behaviour to reduce impacts of development on flora and fauna

Developments are usually associated with increased human presence and this almost always leads to increased interaction with the local flora and fauna. Unfortunately, human behaviour is one of the main causes of nature degradation across the world (Padua 2010). Even the seemingly innocuous act of feeding native fauna at a park can make animals dependent on people and potentially even cause changes in species composition (Orams 2002; Green and Giese 2004).

Measures to modify human behaviours that can harm wildlife are described in this chapter. Needless to say, human behaviour is a highly variable factor and, because of this, we do not attempt to account for all actions that happen at development sites. Instead, it is assumed that a development has a moderate standard of management in regard to people–wildlife interactions and people do not cause wilful harm to wildlife. The measures discussed in this chapter are commonly used, but, because the measures are seemingly simple, they can be taken for granted and poorly executed.

Education and effective communication are important tools that can bring about positive changes in human behaviour. People are often more willing to change their behaviour once they understand the causal relationship between their action and its effect on wildlife. Harm to wildlife often occurs simply because people do not understand the consequence of their actions or because they do not understand the inherent value of the wildlife concerned.

Environmental professionals have a responsibility to help people understand not only the importance of wildlife but also how individuals can change their behaviour to minimise their own impacts on wildlife. One way of achieving this is by better communicating how to co-exist with wildlife by simplifying conservation science for the public. An example of this is an easy-to-understand educational sign as shown in Figure 5.1.

On the other hand, not everyone values wildlife conservation. Some people are fully aware that their actions are harmful. It is still possible to convince these people to minimise their impacts on wildlife by making them aware that their actions may also affect them personally. These people are likely to be more motivated by monetary fines or by finding out that they could become injured because of their actions.

Importantly, by changing human behaviour, other more elaborate or expensive measures may not be required. For example, lowering the vehicle speed limit in ecologically sensitive areas would be less expensive, and less disruptive to animal movement, than fencing both sides of a road to prevent collisions with fauna. Because collisions between vehicles and fauna are so

Figure 5.1. Wildlife education sign installed at Sydney Olympic Park, New South Wales (Photo: J Gleeson)

prevalent (many thousands of collisions occur each year in Australia; Rowden *et al.* 2008; Ramp and Roger 2008), this chapter dedicates two sections to modifying human behaviour to prevent collisions.

Wildlife education

Overview	
What:	Educating people so that they are better informed to make decisions that do not affect wildlife.
Where:	All developments.
Pros:	People are often more willing to change their behaviour once they understand the causal relationship between their action and its effect on wildlife.
Cons:	There will always be some people who refuse to make their behaviour more wildlife-friendly.
Cost:	Low (e.g. posters)–high (e.g. television advertisements).
Companion measures:	Wildlife education can augment nearly all of the measures included in this book.
Alternative measures:	Fences can be used to separate people from wildlife in some situations (Chapter 7); impacts can be avoided (e.g. by limiting development in or near ecologically sensitive areas) (Chapter 4).

Description and application

A fundamental reason for educating people about wildlife is to influence them to minimise their impact on flora and fauna. People need to understand the consequences of their decisions so that they are better informed when making them. Generally, this is done by raising awareness of wildlife and appealing to people to consider their behaviour. Education is a particularly powerful conservation tool because everyone can participate in wildlife education activities at all types of development.

People are not always aware of the full extent of their impact on wildlife. For example, a speeding motorist may strike a bird causing it to die. The motorist is aware that his decision to speed has caused the death of one bird. However, this motorist does not realise that other speeding motorists have also hit birds along the same road and the local population of birds is now in danger of being lost. In this example, each individual driver may regret hitting a single bird, but the full impact of their collective actions on the bird population is not realised by any of them. If the speeding motorists had been educated to understand that this behaviour would lead to the loss of the local population of birds along the stretch of road, and the conservation significance of this, they might have been more careful while driving.

Many people have an affinity for plants and animals and are genuinely interested in conserving local wildlife (e.g. Miller 2003; FitzGibbon and Jones 2006). Education is a means of empowering someone so that they are able to reduce impacts on wildlife. Sometimes people believe that their behaviour is environmentally sound, but in fact their actions are counterproductive or not as beneficial as they could be. The authors have met livestock graziers who consider themselves environmentally conscious for retaining standing trees. There is no question that this is an environmentally positive action; however, their livestock were permitted under the trees, resulting in loss of the native understorey, localised erosion and suppression of future tree regrowth.

Wildlife education should rarely be thought of as a one-way transmission of information from one party to another. Discussions can enlighten both parties and the creation of relationships encourages voluntary exchange of information (Goldney et al. 1996). For example, the same grazier mentioned above had intimate knowledge of the local environment and was keen to share sightings of the local fauna that he had made over many years in the area. This kind of local knowledge is rarely considered in ecological impact assessments, but a more communicative and collaborative relationship with surrounding landholders regarding the ecology of an area could help gain valuable information that may help to reduce impacts on wildlife. One way that knowledge can be shared is by inviting the exchange of information (e.g. setting up a forum for discussion and action in association with local natural history and landcare groups).

Wildlife education can help to reduce the impacts of a variety of development types and can augment nearly all of the other measures included in this book. Wildlife education is particularly useful to raise awareness of measures that have been put in place at a development site to protect wildlife. For example, people who have been told why nest boxes have been installed around a particular development site, and why they should not be interfered with, are less likely to disturb the animals that use them. People can also be educated on how they should behave in relation to local wildlife at a development site. For example, people can be asked to refrain from feeding wildlife (Figure 5.2) and from entering sensitive habitat areas (e.g. breeding sites).

Educating people at a development site about wildlife can help people become more aware about their local environment. For example, one of the authors visited a development where an employee took a keen interest in reptiles. The employee would photograph the different reptiles

Figure 5.2.   Education sign installed at the Australian Botanic Garden, Mount Annan, New South Wales (Photo: J Gleeson)

that he saw around the development and share them with other site personnel. This built a general community awareness of local reptiles.

A simple conceptual framework that can be used for educating people on how to minimise harm to wildlife is provided in Figure 5.3. The first step involves working out the message that needs to be conveyed. This involves identifying the environmental factor ('factor'), why the potential problem is relevant to the audience ('risk') and how they can avoid the risk ('action') (Figure 5.3).

The second step is to choose a medium, or combination of media, that is most suitable to convey the message. People can be educated about wildlife via various verbal and non-verbal media, as outlined in Table 5.1. The ways people communicate are changing and modern communication methods (e.g. short message services – SMS) are worth investigating.

Some of the media listed in Table 5.1 are dependent on people voluntarily reading the message (e.g. wildlife education signs, brochures, posters, email and SMS). Of these, the more visually appealing media, such as wildlife education signs and posters, help capture people's attention. Multiple page brochures could be used to provide a detailed message to reduce the likelihood that a complex message will be misinterpreted, whereas presentations allow information to be delivered in an engaging way, which increases take up and recall of the message.

Effectiveness

Market research or consultation can be undertaken to improve the likelihood that a message will be well-received (e.g. FitzGibbon and Jones 2006; Dowle and Deane 2009). This involves consulting with a sample of the target audience early in the process to determine their knowledge and misconceptions about an issue. Also understanding the interests, beliefs, experiences

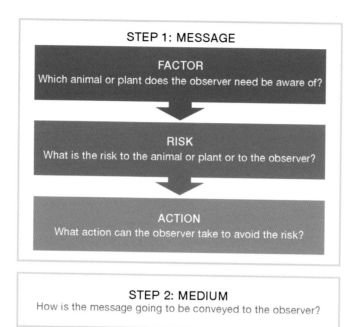

Figure 5.3. Framework for educating people on how to reduce their impacts on wildlife (Diagram: J Toich)

Table 5.1. Media for wildlife education

Medium	Example
Engaging people in discussions	Personnel at a development site can be encouraged to have conversations about wildlife by having a social function to raise money for wildlife conservation.
Engaging people in actions	Personnel at a development site could participate in conservation actions (e.g. tree planting).
Wildlife education signs	Signs can be erected next to a tree to explain its conservation significance.
Posters	Posters with glossy photographs of wildlife displayed in the lunch room can capture the interest of personnel.
Brochures	Brochures with detailed information about what wildlife may be seen and what to do if an injured animal is found can be made available at roadside rest areas for people to take with them to refer back to if they spot an animal that has been injured by a vehicle.
Presentations	Presentations can be tailored to give specific instructions to visitors to a development site so that they do not inadvertently interfere with any conservation initiatives.
Broadcast media	Television or radio advertisements to encourage motorists to slow down to prevent hitting wildlife would reach a large audience.
Email	An email requesting that sightings of a particularly invasive weed be reported can easily be sent to all personnel.
Short message service (SMS)	Messages could be sent via SMS to alert relevant personnel about new access restrictions to part of a development site.
Mobile phone applications 'Apps'	Motorists could elect to be warned when approaching a stretch of road that is prone to vehicle–wildlife collisions.

and attitudes of the target audience can help improve the effectiveness of wildlife education. For example, Ballantyne and Hughes (2006) tested draft wildlife education signs *'in situ'* with a sample of the target audience to ensure that the design and content was persuasive prior to installation.

It is easier for people to grasp concepts when they are provided information that gradually builds upon their existing knowledge base. Complex or highly technical pieces of information should be introduced only once the basis is understood.

Messages should be conveyed in simple and easy-to-understand language regardless of the inherent level of knowledge of the audience. Anyone can be involved in wildlife education; however, speaker qualities are an important factor for making a presentation more memorable for the audience (Knapp and Yang 2002).

Emphasising the personal significance of the message to the audience is likely to raise their interest. For example, a pastoralist may be more inclined to retain habitat for birds in an agricultural setting if he/she knew that birds forage on many agricultural pests.

Educating people about local threatened species can be a good catalyst for raising awareness of wildlife because members of the public are particularly interested in threatened species, due to their seemingly more important value. Sharing interesting or unusual facts about local wildlife can also create interest. For example, platypus (*Ornithorhynchus anatinus*) forage underwater with their eyes closed and, using their bills, they detect an electric field associated with small living prey items. This is an engaging and interesting fact for many people.

Wildlife education is likely to be more effective when a combination of education techniques are used. For example, a sign warning people to stay away from a sensitive habitat area at a development site is likely to be more effective if personnel or visitors entering the site are informed of the existence of the sign and its relevance at an induction.

Fauna crossing signs

Overview	
What:	Fauna crossing signs are signs that are erected along roads to alert motorists that animals might enter the roadway.
Where:	Useful along stretches of road prone to road-kill or along roads that pass through ecologically sensitive areas.
Pros:	Alerts motorists of the need to modify their driving behaviour to prevent a collision with an animal.
Cons:	Effectiveness depends on the willingness of motorists to modify their behaviour.
Cost:	Low.
Companion measures:	Reducing vehicle speed (Chapter 5); reducing attractive habitat along roadsides (Chapter 7); canopy bridges and glide poles (Chapter 8).
Alternative measures:	Avoid impacts (e.g. by alternative road design or location) (Chapter 4). Land bridges and fauna underpasses (Chapter 8) and fauna exclusion fences (Chapter 7)

Description and application

Fauna crossing signs are erected along roads to alert motorists that animals may enter the roadway and to motivate them to modify their driving behaviour to reduce the likelihood of a

Figure 5.4.   Classic Australian fauna crossing sign (Photo: D Gleeson)

collision. It is likely that fauna crossing signs were originally installed to protect motorists, although they are now increasingly being used to conserve animals. Fauna crossing signs are usually installed for highly mobile animals that are likely to exit the road if drivers slow or change lanes (e.g. kangaroos and koalas). They are generally not suitable for small and slow-moving species (e.g. some reptiles and frogs) because, even if a motorist was able to detect a small, slow-moving animal on the road ahead, a collision may be unavoidable unless the motorist stops or swerves from the road. To reduce the impacts of new roads on these types of animals, developers should consider an alternative location for the road or consider incorporating a fauna underpass into its design (Chapter 8).

Standards exist in Australia for the use of signs along public roads (e.g. size and placement). There is greater flexibility in how the content is displayed. Traditionally, fauna crossing signs in Australia typically depicted a black silhouette of an animal on a diamond-shaped, yellow background (Figure 5.4). More recently, the effectiveness of this classic design has been challenged and many new and interesting designs are emerging, as discussed further below.

A fauna crossing sign should convey a message that is easily understood so that it elicits the desired action from the observer. When viewing a fauna crossing sign, an observer should be able to quickly and easily understand which animal is relevant (i.e. the 'factor'), why the potential problem is relevant to them (i.e. the 'risk') and how they can avoid the risk (i.e. the 'action') (Figure 5.5). This educational framework was introduced earlier in this chapter.

The 'factor' should be easily recognisable to the target audience. Communicating the factor with a picture, rather than words, may be preferable to overcome language barriers (e.g. tourists). This doesn't always work, however. Some tourists have mistaken the image of a black silhouette of an animal on a yellow background as tourist information about areas for viewing fauna (Magnus *et al.* 2004). Better designed signs use easy-to-understand language, universal symbols and provide additional information regarding risk and action to prevent such misunderstandings. The length of the section of road applicable to the fauna crossing sign is also often added to the sign as part of the 'factor' so that motorists are aware that they should be cautious for a specified distance after the sign (Figure 5.6).

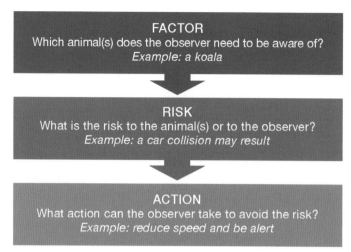

Figure 5.5.   Fauna crossing signs should be designed to be simple and quick to understand by passing motorists. This can be achieved by incorporating the 'factor', 'risk' and 'action' into the design (Diagram: J Toich)

Signs that clearly represent the 'risk' to the fauna and/or the observer can motivate the observer to take action (Figure 5.7). A sign depicting an animal colliding with a car or an injured animal may be appropriate. The 'action' the observer should take to reduce the risk of the problem occurring must be immediately obvious, particularly because motorists generally pass by the signs at relatively high speeds. An example of an 'action' that could be incorporated into a fauna crossing sign is to reduce vehicle speeds by a specified amount (as discussed later in this chapter). Another 'action' may be to request drivers to report injured animals or an incident of road-kill.

Fauna crossing signs are generally more suitable for use along long segments of roads with a high incidence of road-kill (Hobday and Minstrell 2008). Other measures may be more appropriate for more localised high-risk areas, such as: alternative road design or location (Chapter 4); reducing traffic speed (this chapter); land bridges (Chapter 8); fauna underpasses (Chapter 8); canopy bridges (Chapter 8) or glide poles (Chapter 8).

Effectiveness

Fauna crossing signs will not eliminate the risk that roads pose to fauna, but well-designed signs can motivate some motorists to modify their behaviour and this will prevent some

Figure 5.6.   (a) Koala walking (Photo: J Gleeson) compared with (b) a koala sitting in tree (Photo: D Gleeson)

Figure 5.7.   An eye-catching modification to a sign in Kuranda, far north Queensland that clearly shows the 'risk' to the southern cassowary (*Casuarius casuarius johnsonii*) (Photo: M Gleeson)

collisions with animals. Factors likely to influence the effectiveness of fauna crossing signs include design and placement. These factors are discussed below.

Design

Studies have found that the black silhouette of an animal on a diamond-shaped, yellow background (e.g. Figure 5.4) is poor at reducing road-kill in Australia (Coulson 1982; Gardyne 1995). Standard fauna crossing signs have also been reported as largely ineffective in Europe because motorists become desensitised to them (Bank *et al.* 2002).

The effectiveness of fauna crossing signs is likely to be enhanced by modifications that command human attention such as:

- adding a speed limit (this chapter)
- adding flashing lights (Bank *et al.* 2002)
- using fibre optics to display a reduced speed limit (Bank *et al.* 2002) or flashing lights (Gordon *et al.* 2004) when triggered by animal presence
- making signs physically larger (Al-Ghamdi and AlGadhi 2004)
- adding diamond reflective material (Al-Ghamdi and AlGadhi 2004)
- using portable, rather than permanent, signs (Hardy *et al.* 2006) so that signs can be used at the most relevant times (e.g. seasonally for migratory animals).

If a black silhouette of an animal is to be incorporated into a sign, the animal would best be depicted as they would be found in the risk situation (Queensland Department of Main Roads 2000). For example, a koala walking (Figure 5.6a) may be more effective than the representation of a koala passively sitting in a tree (Figure 5.6b).

Some more recent fauna crossing signs display road-kill statistics for an upcoming stretch of road that motorists are about to travel to demonstrate the importance of slowing down. It is vital to maintain up-to-date fatality statistics so that people do not lose faith in the sign.

Figure 5.8   Modern electronic road signage used to catch the attention of motorists and provide them with a tailored message. This sign is in Cleveland, Queensland (Photo: J Gleeson)

Electronic road signage, such as the sign shown in Figure 5.8, are programmable, which enables a tailored message to be displayed and updated over time.

'Animal detection systems' that use sensors to detect large mammals (e.g. deer) as they approach the road have been used overseas in North America and Europe to activate signs that urge motorists to reduce their travelling speed and be more alert (Huijser *et al.* 2010). These systems can decrease wildlife–vehicle collisions (see review in Huijser *et al.* 2010).

Education is likely to play a large role in the effectiveness of fauna crossing signs. Supplementary information at rest stops that explains why fauna crossing signs have been installed along the road may encourage motorists to pay attention to the signs while driving. Such information could include pictures of road-kill and/or car accidents caused by collision with fauna to further motivate drivers to take care.

Placement

A fauna crossing sign should be placed in a location that ensures they are seen and provides ample response time for motorists. The effectiveness of fauna crossing signs can depend on how many are erected. If there are too many signs, observers are likely to become desensitised and consequently ignore the message. If signs are installed in areas that only present a low risk to animals, people are likely to become complacent and the sign may lose its effect when used in high-risk areas. Fauna crossing signs should be reserved for areas that regularly pose hazards. This can be determined by local knowledge, site-specific studies or monitoring. As mentioned earlier, fauna crossing signs are more suitable for use along long segments of roads with a high incidence of road-kill (Hobday and Minstrell 2008).

Impacts on other aspects of the environment

A small area of ground disturbance is required for the physical placement of fauna crossing signs. There is normally some flexibility where they are placed so particular habitat features or significant vegetation can usually be avoided.

Reducing vehicle speed

Overview	
What:	Lowering vehicle speed so that motorists have a longer response time to avoid collision with fauna.
Where:	Useful where a road presents a high risk to animals, which are likely to survive if motorists reduce their speed.
Pros:	May prevent fatality of fauna.
Cons:	Measures that depend on motorists choosing to modify their driving behaviour (e.g. recommended lower speed limits) may not be very effective.
Cost:	Low (e.g. signs)–medium (e.g. traffic-calming structures).
Companion measures:	Fauna crossing signs (this chapter); canopy bridges and glide poles (Chapter 8) or reducing attractive habitat along roadsides (Chapter 7).
Alternative measures:	Avoid impacts by using an alternative road design or location (Chapter 4), erecting fauna crossing signs (this chapter) or reducing attractive habitat along roadsides (Chapter 7). Land bridges and fauna underpasses (Chapter 8) and fauna exclusion fences (Chapter 7).

Description and application

At higher vehicle speeds, motorist response time is reduced and this increases the likelihood that fauna will be hit. The speed of vehicles in a given area can be reduced by recommending a lower travelling speed, enforcing a legal speed limit or incorporating traffic-calming structures (also called 'slow points') into the road design. Measures to reduce traffic speed are not only relevant to public roads; they are also applicable to private roads in development sites because vehicles travelling along these roads can also pose a risk to fauna. Reducing vehicle speed aims to give animals adequate time to move out of the path of an oncoming vehicle, as well as to give the motorist more time to respond to the situation.

Vehicle speed limits

Signs that specify lower speed limits are usually installed along roads primarily for human safety, but can also be used with the intent of preserving animals. Speed limits can be either recommended or legally enforced. The threshold at which speed causes animal–vehicle collisions will vary depending on a number of factors such as the curvature of the road, topography or presence of adjacent habitat. Existing roads can easily be retrofitted to include signs that specify lower speed limits.

Traffic-calming structures

Traffic-calming structures are installed on roads in order to compel motorists to lower their travelling speed (Figure 5.9). Installing these structures can lead to less traffic overall because some motorists prefer to take an easier alternative route. Inclusion of traffic-calming structures at the planning stage of road design is generally more economical than retrofitting existing roads.

Effectiveness

Traffic-calming structures can reduce the risk of vehicle–fauna collisions occurring (Jones 2000; van Langevelde and Jaarsma 2009) (Box 5.1). The inclusion of structures such as speed bumps and traffic obstacles is likely to be more effective in reducing animal–vehicle collisions than erecting signs that recommend lower traffic speeds. Motorists can choose to ignore speed

Figure 5.9.   A series of traffic-calming structures designed to compel motorists to lower their travelling speed to prevent collisions with fauna have been incorporated into the recent upgrading of Gap Creek Road, Brisbane, Queensland (Photo: J Gleeson)

recommendations but are compelled to lower their travelling speed in response to traffic-calming structures. This idea is supported by a study undertaken by Ramp *et al.* (2006), which revealed legal speed limits on roads within Royal National Park in New South Wales were not adhered to, particularly at night when most animals were using the road. They suggested that a simple reduction of speed limit in these locations may not solve the problem and suggested that other measures such as traffic obstacles would be needed to guarantee that motorists would reduce their travelling speed.

These findings by Ramp *et al.* (2006) are similar to those of the 'Koala Speed Zone Trial' which was conducted on the 'Koala Coast', South-East Queensland by Dique *et al.* (2003). The aim of the trial was to assess the effect of differential speed limit signs (60 km/h at night between August to December and 80 km/h at all other times) on the number of koalas hit by vehicles. Koalas are often at risk as they come to ground to move between suitable habitats (Figure 5.10). They found that vehicle speeds did not decrease despite signage mandating lower speed limits and, not surprisingly, there was no reduction in the number of koalas hit. On the other hand, an earlier study by Gardyne (1995 in Queensland Department of Main Roads 2000) found that lowering the speed limit from 100 km/hour to 80 km/h reduced the number of koalas killed.

The behaviour of the animals in the vicinity of a particular road should be taken into consideration when determining suitable traffic speeds. For example, if the animals most often observed as, or most likely to become, road-kill are crepuscular/nocturnal (e.g. most Australian mammalian fauna) reducing speed limits at dawn and dusk only is likely to be adequate.

Box 5.1.   Impacts of traffic speed on local populations near Cradle Mountain, Tasmania

The negative impact of road mortality on local fauna populations, as well as the positive effects of implementing responsive mitigation measures, was observed near Cradle Mountain in Tasmania. Jones (2000) was part way through an ecological study when a tourist road into the northern end of the Cradle Mountain–Lake St Clair National Park was widened and sealed in June 1991. Over 17 months, Jones (2000) observed that the resident population of 19 eastern quolls (*Dasyurus viverrinus*) became extinct and the Tasmanian devil (*Sarcophilus harrisii*) population, consisting of 39 individuals, halved. This observation coincided with a dramatic increase in the number of road-kills.

The negative impact on eastern quoll and Tasmanian devil populations was attributed to an increase in modal speed of approximately 20 km/h and a greater increase in maximum speed. Traffic-calming structures were installed to slow vehicle speeds by 20 km/h. Other measures were introduced to reduce the incidence of road mortality, including increasing motorist awareness by erecting signs and distributing pamphlets, installing reflectors on the roadside to deter animals from crossing the road and by providing escape opportunities from the road.

Within 2 years following implementation of the measures, Jones (2000) observed that the eastern quoll population had recovered to 50% of its former size and there was some indication that Tasmanian devil populations were also recovering. The combined measures appeared to have successfully reduced the impact of traffic speed on fauna, but it was not possible to determine the effectiveness of individual measures.

Figure 5.10.   A koala on the ground near a roadway in South-East Queensland (Photo: J Gleeson)

The colour of the fur of an animal can influence how easily motorists can detect them and avoid collision. Hobday (2010) found that light-coloured animals such as European hares (*Lepus capensis*), Tasmanian bettongs (*Bettongia gaimardi*) and eastern barred bandicoots (*Perameles gunnii*) are more easily detected than larger but darker species such as Tasmanian devils (*Sarcophilus harrisi*) and eastern quolls (*Dasyurus viverrinus*). Hobday (2010) suggested that vehicle speeds would need to be lowered from 100 km/h to 50 km/h if motorists are to have enough stopping time after viewing a Tasmanian devil on the road.

Placement

Identification of areas where animal–vehicle collisions are most likely to occur (i.e. 'hotspots' or sometimes called 'black spots') is an important first step towards mitigating the fatalities of animals on roads (Ramp *et al.* 2005a). Hotspots can be identified as areas where road-kill is most often observed or areas that are known to be frequently used by animals. Measures to reduce fauna fatalities, such as lowering traffic speed, are likely to be substantially more effective when concentrated in hotspots. In recent years, modelling approaches have been used to identify hotspots (e.g. along the Snowy Mountain Highway in southern New South Wales; Ramp *et al.* 2005a). Modelling enables fatality hotspots for different fauna species in different regions to be predicted.

Whether a traffic-calming structure will be effective in maintaining a species in a landscape depends on the size of the traffic-calmed area as well as the area of available habitat either side of the road and its carrying capacity (van Langevelde and Jaarsma 2009).

Impacts on other aspects of the environment

Signs have little adverse impacts on other environmental aspects because only a small area of ground disturbance is required for the physical placement of these structures. It is likely that there is some flexibly on the placement of signs, so particular habitat features or significant vegetation can be avoided. Traffic obstacles (slow points) would require widening of the road at points and therefore slightly more land clearance.

6

Measures to reduce direct fauna and flora mortality during land clearance

Land clearing is commonly required to facilitate development in Australia and overseas. Land clearance can include the removal of native vegetation, movement of terrain and/or modification of aquatic systems. These activities usually result in a loss of flora and fauna habitat as well as physical trauma (e.g. injuries or fatalities) to animals using the habitat at the time. Less agile, hibernating and nesting animals are particularly vulnerable.

Before deciding to undertake land clearance activities, the feasibility of using measures to avoid impacts on wildlife should be investigated (Chapter 4): in particular, relocating, redesigning or at least staging the development. Although clear-fell vegetation clearance (i.e. an entire stand of vegetation is removed in one felling) is still prevalent (Figure 6.1), more

Figure 6.1. An excavator clearing regrowth vegetation in the Hunter Valley, New South Wales (Photo: J Gleeson)

sustainable practices aimed at minimising impacts should be used preferentially, including restricted clearing patterns, retaining adjacent vegetated areas and selective clearance. For example, removing large trees while retaining a shrub layer beneath powerlines helps avoid unnecessary loss of native understorey plants (Clarke and White 2008).

When land clearance is intended to be only temporary (e.g. mineral exploration), the prospect that vegetation will regenerate following disturbance can be improved by preserving plant root systems and maintaining viable soil seed banks (White *et al.* 1996). The root systems of some plants can be preserved by slashing or folding vegetation rather than removing the entire plant. Even when entire plants are removed, maintaining viable soil seed banks can increase the resilience of the land. Minimising soil disturbance can reduce impacts on native seed banks. In open-cut mining developments, topsoil with a viable native seed bank is usually stockpiled for later use in rehabilitation of post-mine landforms.

Two measures – fauna rescue and plant salvage – can both be used to reduce impacts on fauna and flora during land clearance and are described in this chapter. In some situations, it is also worthwhile salvaging habitat features (Chapter 10) and collecting seed for revegetation (Chapter 10) or preservation (Chapter 11).

Fauna rescue

Overview	
What:	Actions that minimise injury or death of fauna during land clearance activities.
Where:	Beneficial when native fauna are likely to occupy habitat designated for clearing and when suitable habitat and resources are made available elsewhere.
Pros:	Minimises the number of animal injuries and deaths during land clearance and subsequently may help maintain local fauna populations.
Cons:	May lead to greater competition between individuals for limited resources in the remaining habitat.
Cost:	Low–medium.
Companion measures:	Measures to provide additional habitat for flora and fauna impacted by development (Chapter 10); discouraging exotic vertebrate fauna (Chapter 9).
Alternative measures:	Avoiding habitat clearance (Chapter 4); translocation of fauna (Chapter 11).

Description and potential applications

Fauna rescue is undertaken before, and during, land clearance activities to minimise injury or death of individuals. Fauna rescue is worth considering, and is sometimes mandatory, because land clearance can reduce the viability of local fauna populations and increase the risk of local extinctions. Fauna rescue is routinely undertaken by various infrastructure and mining companies in Australia.

Fauna rescue is needed if a development plans to clear fauna habitat or otherwise present a risk to fauna (e.g. trenching for pipeline installation). Impacts on fauna habitat can be avoided using the various measures outlined in Chapter 4, such as relocating or minimising the spatial footprint of a development. Key components of fauna rescue include a pre-clearance fauna survey to gain an understanding of the fauna present and minimising animal injury and fatality during land clearance as described below.

Pre-clearance fauna survey

Ecologists or wildlife handlers generally conduct pre-clearance fauna surveys and, in this role, they are commonly referred to as 'fauna spotters/catchers'. When designing pre-clearance fauna surveys, it is important to consider first the types of animals that could occur based on the type of habitat present. Those animals can then be targeted by using specific survey techniques. Typically, pre-clearance surveys are aimed at birds, bats and/or arboreal mammals, although other faunal groups can be incorporated (e.g. reptiles). Generally, amphibians and other semi-aquatic fauna are the most challenging vertebrate groups to include in a pre-clearance survey program successfully due to their highly specialised habitat requirements.

During pre-clearance surveys, particular attention is paid to habitat features used by fauna because this is where some target animals are likely to be residing. Habitat features can include habitat trees, other forms of vegetation, logs, rock piles, caves, burrows and dens. Habitat trees are living or dead plants that contain habitat features (e.g. hollows for birds or arboreal animals, shedding bark suitable for sheltering bats or reptiles, and occupied nests). Habitat features are clearly identified (e.g. by using flagging tape, fencing, roping or spray paint) so that the feature is not inadvertently disturbed. A global positioning system (GPS) is often used to register the position of the habitat feature for planning purposes and help in locating the habitat feature later.

The survey techniques to locate fauna during pre-clearance surveys can include searching for nests, spotlighting for nocturnal animals, watching hollows after sunset and daytime inspection of hollows using fibre-optics. Observations of secondary evidence (e.g. tree scratches, scats, remains of prey or tracks) are also useful indications that fauna could be present.

Over the past decade, significant technological advances have been made in equipment that can detect fauna presence during pre-clearance surveys. These include bat ultrasound call detectors to survey potential bat roosts, motion video cameras for surveillance of habitat and infrared live feed video to survey habitat features such as hollows, burrows or dens. Radio-tracking has been used to gain an understanding of habitat use and movement of animals and can be used to track relocated individuals.

Habitat structure and terrain will influence the scope of the pre-clearance surveys and the method of vegetation clearance. More open habitats are easier to access and visually survey from the ground.

Minimising fauna injury and fatality during land clearance

Various types of machinery are used to clear land (e.g. excavators, bulldozers and chainsaws), each with their own mode of operation and corresponding level of control when clearing. To minimise the impact to fauna, it is preferable that different machines are used in particular situations. Bulldozers are usually preferable if habitat trees identified as likely to contain fauna need to be felled, especially if an experienced bulldozer operator is able to safely lower the tree slowly because this would result in less disturbance to its inhabitants. The authors have observed bulldozer operators doing this by pushing against the tree near the base and sliding the bulldozer blade down the trunk as the tree is lowered (Figure 6.2). Excavators are generally more efficient at digging or clearing large areas of shrubby vegetation (Figure 6.1).

Maintaining habitat linkages during clearance enables some animals to move away from the clearance activities on their own accord. This can be facilitated by selective or progressive clearance. Of course, for this to be effective, there would need to be suitable nearby habitat for the animals to move in to.

An understanding of animal behaviour is required for the sound management of fauna during land clearance activities. For example, if a tree is shaken with the bulldozer, a bird roosting on a branch is highly likely to fly to another tree. However, a reptile such as a lace

Figure 6.2.   A bulldozer lowering a habitat tree slowly with the edge of bulldozer blade, Hunter Valley, New South Wales (Photo: J Gleeson)

monitor (*Varanus varius*) is more likely to climb higher up the tree or even climb onto a linking tree, which may also require clearance. Care should be taken not to flush animals into areas where they would be susceptible to other threats (e.g. a busy highway).

An understanding of animal biology is equally important. For example, common wombats (*Vombatus ursinus*) can have multiple den entrances, so blocking one entrance will not necessarily ensure that the wombat cannot re-enter the den. Some common situations that arise during land clearance, and possible solutions to reduce impacts on fauna, are provided in Table 6.1.

Capture and release methods (e.g. Elliot or cage trapping) can be used to remove fauna from harm (Hanger and Nottidge 2009). Any uninjured animals rescued during pre-clearance surveys are usually released in similar adjacent habitat. Stress to caught animals should always be minimised. Note that, in Queensland, laws specify that capture and release methods do not apply to koalas (Box 6.1).

Clearing vegetation surrounding a habitat tree may encourage some animals to vacate on their own accord. However, on the downside, it can also render the animals more susceptible to predation because they may need to come to ground to move into more suitable adjacent habitat. If this is likely to be an issue, predatory exotic vertebrate animals can be controlled at the development site before clearance, but the risk from native predators, such as birds of prey, will remain.

Effectiveness

Pre-clearance fauna surveys can inform management and enable animal injuries and fatalities to be reduced during land clearance. Ultimately, the effectiveness of fauna rescue depends on the biology and behaviour of the target fauna, characteristics of the habitat requiring clearance and the characteristics of nearby habitat that will remain. From a man-

Table 6.1   Common situations that arise during land clearance and possible solutions to reduce impacts on fauna

Situation	Possible solutions
A tree hollow is found to be inhabited by an arboreal mammal	• Wait until the animal has left the hollow on its own accord and block the entrance so they are unable to re-enter the habitat feature before clearance, after determining no young are present. • If it is safe to do so, fell the tree slowly so that the entrance to the hollow remains unobstructed, allowing the animal to leave unaided. • Clear the surrounding vegetation first (only if it is designated to be cleared anyway) to encourage animals to vacate the habitat feature on their own accord. • Apply capture and release methods.
A bird is found sitting on a nest with eggs or young	• Ideally the nest should be left undisturbed until nesting activities have been completed (i.e. young have left the nest). • Carry out clearing in non-breeding season. • Advanced fledglings can be collected and temporarily cared for by a wildlife carer for subsequent release. • As a preventative measure, inactive nests can be removed during pre-clearance surveys to minimise the chance that a new nesting attempt would occur before clearance.
A bird is found roosting in a tree	• Wait until the roosting bird has left the tree. • Safely disturb or shake the tree with a dozer or excavator to encourage the bird to fly to a different tree.
Bats are found roosting in a tree	• See methods for arboreal mammals above. • It is possible to fell a tree and find bats exiting a hollow that was not seen prior to clearance. In this event, the bats can be collected (taking the appropriate precautions to avoid contracting disease from the bats) and kept covered in a box for re-release at dusk or during the night.
A bird nest is observed in a tree but it is unknown whether it is being used or not	• Watch the nest for a period of time to see whether any birds fly to or from the nest. • A cherry-picker (Figure 6.3), crane, ladder or climbing gear can improve visibility. • Safely disturb or shake the tree with a dozer or excavator to check if any birds fly away from the nest. • If it is safe to do so, fell the tree slowly in a way so that the nest is not likely to be crushed by the tree and check the nest for young.
It is suspected that fauna are still within a habitat tree after clearance	• Inspect felled habitat trees to rescue animals that will not move away on their own accord. • Felled trees potentially containing nocturnal fauna can be left overnight prior to removal to allow fauna to leave the hollow unaided.
Once the tree is felled	• Search fallen tree for injured or trapped fauna.

agement point of view, successful fauna rescue requires forward planning to integrate the pre-clearance surveys into development timeframes, competent machine operators and experienced fauna spotter/catchers. It must not be forgotten that the safety of the people involved in fauna rescue, as well as any bystanders, is paramount, so any activities or situations that are unsafe should be avoided.

Despite best efforts to reduce the impact of land clearance, fauna injury or deaths are always a possibility due to the nature of land clearance and because fauna behaviour is not always predictable. For example, one of the authors has observed a pair of rainbow lorikeets (*Trichoglossus haematodus*) unexpectedly occupying what appeared to be a hollowed-out root below

Figure 6.3.   A cherry-picker being used to inspect a bird nest in the top of a tree to determine if young are present (Photo: J Gleeson)

ground level as opposed to their usual habitat of arboreal tree hollows. Animals with minor injuries can be collected for veterinary assessment and temporarily cared for by a wildlife carer.

In Australia, many community-based groups have demonstrated success in the rescue and rehabilitation of sick, injured or orphaned wildlife (e.g. see Royal Society for the Prevention of Cruelty to Animals 2008). The Australian National Guidelines for Animal Welfare (Australian Department of Agriculture, Fisheries and Forestry 2008) state that if rehabilitation of an injured animal is unlikely, or successful release of rehabilitated animals is unlikely and if there are no exceptional conservation reasons or needs to keep the animal in captivity, the animal should be euthanased.

Impacts on other aspects of the environment

Animals that are encouraged to move on or rescued during pre-clearance fauna surveys or land clearance need to be able to find and use habitat resources such as food and shelter in their new home. The surrounding habitat must have sufficient resources to sustain the animals already living in this environment, as well as the new arrivals, otherwise competition can lead to problems (e.g. territory challenges). If their new home is already near maximum carrying capacity or habitat resources are likely to be deficient, measures to provide additional habitat for flora and fauna can be investigated (Chapter 10).

In many situations, the home range of the displaced individuals will extend into the remaining habitat that was not cleared. Individuals of the same species already occupying the remaining habitat are often part of the same population.

There is no suitable adjacent habitat in some heavily developed areas. If avoiding land clearance is not an option, translocation of displaced individuals to distant habitats may be

Box 6.1.   Koala spotting in South-East Queensland

The koala (*Phascolarctos cinereus*) is an iconic Australian species and careful planning is required to ensure its conservation (Figure 6.4). In South-East Queensland, koala populations are under increasing pressure from development. The conservation benefits of pre-clearance surveys for the koala are acknowledged through legislation because a 'koala spotter' is required to be present during clearance of koala habitat trees (Queensland Environmental Protection Agency 2008a, b). The *Nature Conservation (Koala) Conservation Plan 2006* (Queensland Environmental Protection Agency 2008b) stipulates that the role of a koala spotter is surveillance-based and, as such, koalas must located and allowed to move from the development site un-aided before clearing the tree.

Figure 6.4.   Koala, South-East Queensland (Photo: J Gleeson)

investigated (Chapter 11). Fauna rescue generally does not involve translocation activities, such as the introduction of a species outside its historical known range, re-introduction of a species into an area that is part of its historical known range or supplementation of additional individuals into a different population. If these translocation activities are proposed, the appropriate ecological investigations should be undertaken to avoid any unplanned secondary impacts. As described in Chapter 11, there are legal and logistic complexities with translocation of animals to distant habitats.

Plant salvage

Overview	
What:	Removal of select native plants from an area approved for development and planting them somewhere else.
Where:	Useful when plants (particularly rare, slow growing and long-lived plants) would otherwise be lost to development.
Pros:	It may be possible to fast-track revegetation by using salvaged plants.
Cons:	Not suitable for many species of plants because they do not transplant well.
Cost:	Low–high (depending on number of plants).
Companion measures:	Other valuable habitat features can be salvaged as well (Chapter 10); revegetation and restoration of ecosystems (Chapter 10).
Alternative measures:	Avoiding land clearance (Chapter 4); translocation (Chapter 11); seed preservation (Chapter 11).

Description and potential applications

Plants that would normally be destroyed in an area approved for development can sometimes be salvaged and transplanted to another location. This can help reduce the impact of development on wildlife in two ways. Firstly, the plants themselves are saved and may further contribute to the perpetuation of the species. Secondly, the plants can improve habitat opportunities for wildlife if planted in areas previously affected by a development (e.g. a mined area).

Transplanting salvaged plants into adjacent rehabilitating landscapes aims to help maintain the existing local population by retaining plants from the same population. It should be emphasised here that transplantation is discussed in this chapter as the localised movement of plants to an adjacent disturbed area undergoing rehabilitation, not to a natural ecosystem. Hence, this chapter does not discuss plant translocation; instead this is discussed in Chapter 11.

Rehabilitation is not a quick process and, even at its fastest, ecological values equivalent to pre-disturbance vegetation can take hundreds of years to achieve, if at all. Rehabilitation of disturbed areas can be fast-tracked by supplementing traditional methods (e.g. seeding and tubestock) with plants salvaged from development sites. Slow-growing species will take a long time to reach reproductive maturity. Transplanting these plants can have significant ecological benefits. Grass trees (*Xanthorrhoea* spp. from the family Xanthorrhoeaceae) (Figure 6.5), cycads (order Cycadales) as well as *Kingia australis* and *Dasypogon* spp. (from the family Dasypogonaceae) are examples of slow-growing and long-lived plants that can be transplanted. Plants requiring conservation, other than those that are slow growing, have also been salvaged (e.g. tree ferns from the family Cyatheaceae have been transplanted to promote their conservation in Mexico; Eleuterio and Perez-Salicrup 2009).

Figure 6.5. Transplanting a grass tree (*Xanthorrhoea preissii*) at Huntly Mine, Western Australia (Photo: JM Koch)

There is currently a low usage of salvage and transplanting techniques in Australia, but its use is growing in the commercial plant nursery industry. There are specific companies in Australia that specialise in the salvage and resale of slow-growing plants. Approximately 3000 *Xanthorrhoea preissii* grass trees are salvaged per annum by a leading commercial operator in Perth (Lamont *et al.* 2004). Many of these salvaged plants are planted in residential and commercial gardens where they can still provide habitat (e.g. nectar source for birds). Plant salvage and transplanting is an option if the development plan provides areas that require rehabilitation at the same time as other areas being cleared.

Effectiveness

Based on current technologies, it is not viable or efficient to salvage and transplant all plant species and, for many species, the use of seed or tubestock is a more effective and reliable way to revegetate large areas. Effectiveness of transplanting can be considered in regard to its success in replicating natural communities (Fahselt 2007). However, effectiveness is considered here as the successful establishment of the transplanted plant. We recognise that the plant can provide habitat even if it does not replicate a natural community. The plant should be self-sustaining without ongoing intervention (e.g. no long-term supplementary watering). In Western Australia and Queensland, at least two mining companies have established salvaged grass trees from clearance areas ahead of the mining front and transplanted them into adjacent post-mine rehabilitated landforms (Brennan *et al.* 2005) (Figure 6.5).

Transplanting into rehabilitating landscapes does not only help maintain local diversity (Herath and Lamont 2009), but the addition of mature plants can improve ecosystem function. For example, grass trees provide important foraging, shelter and/or nesting habitat resource for mammals, such as the southern bush rat (*Rattus fuscipes*) (Frazer and Petit 2007) and mardo (*Antechinus flavipes leucogaster*) (Swinburn *et al.* 2007). Adding mature plants into rehabilitated landscapes can improve ecosystem resilience. Herath and Lamont (2009) found transplanting mature plants that resprout following fire (e.g. grass trees), can increase the persistence of resprouters in restored sites following fire. Young plants have poorly developed lignotubers that cannot regenerate following fire, potentially leading to the loss of the species in rehabilitating sites.

Whether or not transplantation of a salvaged plant leads to its successful establishment is highly dependent on the characteristics of the particular plant. Transplanting methods must be tailored to meet the needs of subject species. Techniques for effective transplantation of grass trees are now well established and used by commercial plant nurseries.

Impacts on other aspects of the environment

Plant salvage and transplanting as defined here is not likely to have adverse impacts on other environmental aspects. As stated above, if plants are to be transplanted into a natural ecosystem, the activity is likely to constitute translocation, which presents additional risks as described in Chapter 11.

7

Measures to exclude or deter fauna from a development site

This chapter describes measures that are used to exclude or deter fauna from a component of a development that could otherwise harm them. Animals are sometimes physically kept away from a hazardous feature by using barriers such as fences and coverings. Alternatively, animals can be dissuaded from being in the area of a hazard by removing attractive habitat or by using sensory deterrents (e.g. visual, chemical or auditory) to scare them away.

Measures to exclude or deter fauna are used in a diverse range of developments all over the world. They are commonly used to prevent animals from threatening human safety (e.g. preventing collisions between birds and aircraft, which can cause planes to crash) or interfering with the operation of some developments (e.g. aquaculture farms and waste management facilities). The use of measures to exclude or deter fauna to reduce the impacts of *wildlife on development* is outside the scope of this book.

It is always better to avoid impacts altogether by removing a hazard rather than to rely on an exclusion device or deterrent, because such measures require continual maintenance and, particularly with sensory deterrents, the risk to animals is never entirely eliminated. If a component of the development is hazardous, can the development still function without that particular component? Not all hazards are avoidable, or can be reduced, so measures to deter fauna in these situations are essential. For example, deterrence of birds is required at airports to avoid collisions with aircraft (Australian Transport Safety Bureau 2010).

Selecting the most appropriate measure to exclude or deter fauna depends on the species at risk, the risk period and the size and type of hazard. Physical barriers (i.e. fences and coverings) are usually successful at excluding fauna from a development site (Figure 7.1); however, the cost of physical barriers sometimes precludes their use. In these cases, the site manager sometimes tries sensory deterrents to scare fauna away. Many deterrents are used on a trial-by-trial basis without rigorous scientific testing. Experience shows that an integrated system using a combination of measures is often the best way to deter fauna from a development site.

Figure 7.1.   Fauna exclusion fence along Old Cleveland Road, Redland Shire, near Brisbane Queensland (Photo: J Gleeson)

Fauna exclusion fences

Overview	
What:	Fences can physically prevent non-flying fauna from entering an area.
Where:	Useful where a component of development poses a hazard to non-flying fauna.
Pros:	Protects non-flying fauna from the dangers of entering a development site.
Cons:	Possible isolation of individuals or subdivision of populations. Continual maintenance is required to ensure fence integrity.
Cost:	Medium.
Companion measures:	Land bridges and fauna underpasses (Chapter 8).
Alternative measures:	Avoid using a hazard to fauna (Chapter 4); fauna crossing signs and reducing traffic speed (Chapter 5); reducing attractive habitat surrounding the development (this chapter).

Description and potential applications

Fences or barriers can keep non-flying native fauna away from hazardous components of a development. When used in this way, they are called 'fauna exclusion fences' (Figure 7.1). These fences are referred to as 'fauna guiding fences' or 'drift fences' when they are used to guide non-flying fauna to, for example, the entry to a land bridge or a fauna underpass (Chapter 8).

Other types of fences discussed in this book include fences to avoid impacts on vegetation (Chapter 4), fauna inclusion fences (Chapter 8), fences to exclude exotic vertebrate animals (Chapter 9) and fences to assist revegetation (Chapters 8 and 10).

Table 7.1. Different design aspects of fences to exclude different types of animals

Type of animals	Design aspects
Climbers (e.g. possums and koalas)	The top of the fence should prevent animals from climbing over the fence. Designs include top-hat, floppy-top or solid barriers. Top hat is a term used to describe an additional component attached to the top of the fence (typically curved formed metal). Floppy-top is a term used to describe an extension of curved mesh at the top of the fence.
Diggers (e.g. wombats and echidnas)	The base of the fence should prevent animals from digging under. A fence apron is an extension of the fence that is buried at the base of the fence and deters animals from pushing or digging under the fence.
Jumpers (e.g. kangaroos, wallabies and bandicoots)	Fence should be high enough to prevent animals from jumping over. Top-hat, floppy-top or solid barrier designs may prevent animals from jumping over the fence.
Small size (e.g. amphibians and reptiles including turtles)	Material on the lower portion of the fence should prevent animals from passing through the fence. Small size of mesh, reduced strand spacing or solid barrier can prevent animals from passing through.

Fauna exclusion fences have the advantage of providing a high level of confidence that fauna will not be able to access the hazard. Hazards that can be fenced include roads, holes (e.g. wells or mine shafts), storages with hazardous materials, or residential estates with domestic animals such as cats and dogs. Fauna exclusion fences are often used to reduce impacts on ground-dwelling native fauna. Reducing impacts on this faunal group is particularly important in Australia because there are many other threats that put many of these species at risk of extinction (e.g. predatory exotic animals; Short and Smith 1994; Moseby *et al.* 2009).

Fauna exclusion fences are increasingly being incorporated into road designs, especially in eastern Australia. Fauna exclusion fences are often appropriate to use along roads with high traffic volumes or if there is a species of concern that is likely to be impacted (Jaeger and Fahrig 2004). Fauna exclusion fences can be particularly useful on curved stretches of road where there is reduced visibility and motorist stopping time is limited.

The most appropriate design of fence depends on the characteristics of the fauna to be excluded in a given situation, as shown in Table 7.1.

Common types of fauna exclusion fences are listed in Table 7.2 and described below. Solid fences, made from wood or concrete, can also perform a similar function by inhibiting fauna movement, and these types of fences are particularly useful for separating residential areas from natural habitat.

Table 7.2. Common types of exclusion fence

Measure	Description
Floppy-top fences	Chainmesh fence with curved mesh at the top of the fence.
Top-hat fences	Chainmesh fence with a formed metal curve attached to the top of the fence.
Metal-sheeted mesh fences	Chainmesh fence fitted with a metal sheet in the top portion of the fence.
Amphibian fences	Fences of various designs with a curved top. These fences have been made from recycled plastic, concrete, durable rubber or mesh covered with shade cloth.
Temporary fences	Temporary, easily transferable barriers with various designs and materials.

Figure 7.2.   Floppy-top fence showing the metal post curved towards the fauna habitat and the mesh extending above the post, Pacific Highway, Queensland (Photo: D Gleeson)

Floppy-top fence

Floppy-top fences are designed to prevent animals from climbing over the fence. These fences have curved mesh at the top of the fence (the 'floppy-top') that folds down when an animal tries to climb on it (Figure 7.2). The optimal degree of curvature of the floppy-top has not been experimentally tested across a range of animals. Broadly speaking, it should be curved enough that the floppy-top folds down when the animal attempts to climb it but not so much that the animal is able to reach the top of the floppy-top from its base. An angle of 45 degrees (Figure 7.2) has been adopted by industry (New South Wales Roads and Traffic Authority 2011).

These fences can also have a fence apron to exclude digging animals. Smaller fauna that can move through standard size mesh (e.g. amphibians and reptiles) can also be excluded by adding a smaller size mesh, shade cloth, metal, plastic or rubber barrier to the lower portion of the fence. Floppy-top fences have been installed in Queensland and New South Wales to exclude koalas (*Phascolarctos cinereus*) from roads.

Technical drawings for a floppy-top fence are provided by the New South Wales Roads and Traffic Authority (2011). These fences consist of steel posts (1167 mm to the bend of the post and 400 mm past the bend), a minimum of 500 mm overhang of mesh to form the floppy top and an apron of 1000 mm fastened to the ground (New South Wales Roads and Traffic Authority 2011). Galvanised steel chainmesh wire is available in a range of colours, including green, to improve aesthetics.

Top-hat fences or capped fence

Top-hat fences, also known as 'capped fences', are a design that pre-dates the floppy-top fence. These fences were originally used in wildlife sanctuaries. Top-hat fences are used in New Zealand to exclude a range of exotic pests from natural areas (e.g. Karori Sanctuary) (Day and MacGibbon 2007; Connolly *et al.* 2009).

Figure 7.3.   Metal-sheeted mesh fence, Cleveland, Queensland (Photo: J Gleeson)

Top-hat fences feature a formed metal curve attached to the top of a mesh fence to keep out climbing animals (a larger curved metal fence top than that shown in Figure 7.5). These fences are generally a minimum of 1.8 m high with a 250 mm diameter cap and a 400 mm fence apron buried 100 mm deep (Long and Robley 2004). Top-hat fences are the most elaborate fence described in this chapter and are generally more costly to install.

Metal-sheeted mesh fence

As the name suggests, metal-sheeted mesh fences are chainmesh fences fitted with a metal sheet that animals cannot climb (Figure 7.3). These fences can be fitted with a fence apron to exclude digging animals and a lower barrier to exclude smaller fauna that can move through standard size mesh.

Technical drawings for a metal-sheeted mesh fence are available in the *Standard Drawings Roads Manual* (Queensland Government 2009). These fences are around 1.8 m high and a 600 mm metal sheet is attached along the top of the fence. The posts, bracing and supports are installed on the opposite side to the animals to prevent them from climbing.

Metal-sheeted mesh fences can be more cost-effective than floppy-top fences (Queensland Department of Transport and Main Roads 2010). However, a disadvantage is that their flat surface can be targeted by spray-painting vandals, as the authors have observed on those installed along the Bruce Highway, Queensland.

Amphibian fence

Studies have shown that roads are sometimes associated with significant frog mortalities (Goldingay and Taylor 2006; Glista *et al.* 2008; Andrews *et al.* 2008). Amphibian fences are

Figure 7.4.   Amphibian fences at Sydney Olympic Park, New South Wales (Photos: J Gleeson)

viewed as one way to minimise vehicle–frog collisions. Various designs of amphibian fences are used overseas (e.g. concave recycled plastic, concrete, durable rubber and shade cloth fences) but most have not been widely tested in Australia. Many frog species are very good at climbing up vertical surfaces so most amphibian fence designs have incorporated a top hat (horizontal or angled) with a lip to prevent frogs from climbing over the fence.

The amphibian fences installed at Sydney Olympic Park for the green and golden bell frog (*Litoria aurea*) were among the first to be installed in Australia (Figure 7.4). These fences are greater than 1 m high and made from a wire frame covered with ultraviolet-resistant shade cloth, with a horizontal top-hat and lip.

Vegetation should generally be removed within 1 m of an amphibian fence to prevent frogs from using the vegetation to jump over the fence. The advantage of the amphibian fences described here is that they are taller than most adjacent wetland vegetation so frogs are less likely to use the vegetation to cross the fence.

Concave recycled plastic barriers, less than 30 cm in height and designed with a curve at the top, have been used overseas. It is not known whether these barriers are effective for Australian frogs. Even if they are effective, there are likely to be other issues with using these designs in Australia, such as these fences can be destroyed in bushfires (which is also an issue for shade cloth fences). To maintain fence integrity there would be an onerous requirement to maintain the adjacent vegetation at a lower height than the barrier. Due to the low height of these fences, there is also a risk of trampling of the fence by other animals (e.g. livestock). Similar low barriers made from concrete also exist and, while these barriers would require less maintenance and are more impervious to physical disturbances including bushfire, there would still be an ongoing requirement to maintain adjacent vegetation at a lower height than the barrier.

An amphibian fence (approximately 400 mm high with a 45° downward sloping top hat) has been installed along the Tugun Bypass, Queensland (Figure 7.5). A rock aggregate has been used in an attempt to suppress vegetation beside the amphibian fence (Figure 7.5). Initial reports are that this design was effective, but the long-term success of this design is so far unknown because it is relatively new. The Queensland Department of Main Roads are currently trialling a durable rubber barrier (400 mm high with a 45° downward sloping top hat of 155 mm and 30 mm lip) that can be retrofitted to existing fences (Queensland Department of Transport and Main Roads 2010).

Figure 7.5.   Amphibian fence erected along the Tugun Bypass, Tugun, Queensland (Photo: J Gleeson)

Temporary fauna exclusion fences

Not all hazards are permanent and temporary fauna exclusion fences are sufficient to protect fauna from temporary hazards. For example, such fences have been used to protect frogs from a V8 Supercar Event at Sydney Olympic Park (White 2008) and to redirect animals flushed onto a road during sugar cane burning in Queensland to safety (Queensland Department of Main Roads 2010). Temporary fencing is often made from materials that will deteriorate over time, so it is important to consider the length of time that a temporary fence is required. These fences may also be erected before land clearance to reduce fauna entering an area proposed to be cleared (Hanger and Nottidge 2009).

Effectiveness

Fauna exclusion fences can be specifically designed to provide a high level of confidence that target fauna will not be able to access a hazard. Bond and Jones (2008) surveyed large fauna road-kill along Compton Road in southern Brisbane before and after installation of metal-sheeted mesh fence on either side of the road. They found that incidences of road-kill was markedly reduced following installation of the fence. The only large road-kill observed during the 4 months after installation were two red-necked wallabies (*Macropus rufogriseus*) that passed through breaches in the fence made by vandals and a wood duck (*Chenonetta jubata*) that appeared to have flown onto the road (Bond and Jones 2008).

Studies on the effectiveness of fencing to control feral animals can give us an idea of how successfully these fences would exclude native fauna with similar traits. Studies by Moseby and Read (2006) and Robley *et al.* (2007) found that some of the agile climbers (cats and foxes) could not scale a 1.8 m fence with a 600 mm floppy-top. Top-hat fences have been used around sanctuary areas to successfully exclude a range of feral animals in New Zealand including rats, possums and rabbits (Day and MacGibbon 2007; Connolly *et al.* 2009).

There are various other designs of fauna exclusion fences, not described in detail in this book, that have been used successfully overseas, such as wire fences (greater than 15 cm mesh size) to exclude white-tailed deer (*Odocoileus virginianus*) (Cleavenger *et al.* 2001).

Factors likely to influence the effectiveness of fauna exclusion fences include its design, placement and maintenance. These factors are discussed below.

Design

As explained earlier, the most appropriate fence design to use for a given situation depends on the biology of the animals to be excluded by the fence. It is recommended that trials are undertaken if it is uncertain whether the fence would work in a given situation. Temporary fences can be erected in order to test the effectiveness of a design before permanent, more expensive structures are installed.

Measures to promote fauna movement across roads (e.g. land bridges) are often used in conjunction with fauna exclusion fences (Chapter 8). Indeed, fauna exclusion fences are often necessary for species that inherently avoid confined passages. In these cases, the terminal ends of the fence should be angled away from the road to guide fauna back into the habitat and reduce the chance that they will move around the end of the fence towards the road.

Placement

The effectiveness of a fauna exclusion fence can be considerably affected by extreme environmental conditions (e.g. flooding, fire or coastal salt spray). If this is likely to be an issue, frequent fence inspections and maintenance will be important (see below). In severe situations, a fauna exclusion fence may need to be relocated or reconsidered.

Where a road passes through fauna habitat, any fauna exclusion fencing is best used on both sides of the road. Fencing just one side of the road can cause fauna to become trapped on the road as fauna from the unfenced side try to cross the road. Understanding the movement routes taken by a species and the location of key habitat resources will provide an insight into optimal fence placement and how species are likely to interact with the fence. Many species have unpredictable movement patterns that can pose a challenge to assessing effective fence placement.

If fencing is deemed inappropriate for a particular location, an alternative measure to deter fauna may be to reduce attractive habitat surrounding the development (discussed later in this chapter). For example, Klocker *et al.* (2006) studied the frequency and causes of kangaroo–vehicle collisions and suggests that roadside vegetation can be altered to make it less attractive for foraging kangaroos (e.g. decrease abundance of grass). Reducing habitat either side of the road could also reduce the chance of striking fauna species that have an inherent avoidance of wide road corridors.

Maintenance

Choosing high-quality fencing materials that do not rust or deteriorate quickly is necessary to ensure fence longevity and to minimise maintenance requirements. The perimeter of fence installed around a hazard can be kept to a minimum to reduce maintenance costs.

The effectiveness of a fauna exclusion fence is reliant on regular inspections for breaches (e.g. Figure 7.6) and their quick repair (Puky 2005), because some fauna can rapidly exploit these breaches. Bond and Jones (2008) recorded road-kill within 12 hours of breaches in a fence being noticed. Fauna with designated movement paths can also try to break through the fence (e.g. wombats). In these situations, it may be appropriate to maintain the pathway (e.g. by leaving a large section of a road unfenced) or installing land bridges or underpasses (Chapter 8).

Figure 7.6.   Floppy-top fence that requires repair due to damage: a fallen tree branch, Pacific Highway, New South Wales (Photo: J Gleeson)

Ongoing and frequent management of adjacent vegetation is important for maintaining an effective fauna exclusion fence because animals can use overhanging vegetation to cross the fence. No vegetation should grow on or over the fence, because small mammals, snakes and tree frogs have been observed using vegetation growing on barrier walls to gain access to roadways (e.g. Dodd *et al.* 2004). As an example, the Queensland Department of Main Roads (2010) specifies a 3 m buffer that should be free from vegetation (excluding grasses) for floppy-top fences erected to exclude koalas. Management of adjacent vegetation can also be particularly important for fence effectiveness in bushfire-prone areas.

Impacts on other aspects of the environment

Fauna exclusion fences that are used inappropriately can pose a threat to biodiversity. Barriers to movement created by fauna exclusion fences erected along roads can sometimes have a greater impact on a species than the collisions with vehicles that would have occurred if the road had remained unfenced. Fences, particularly when used in long stretches, can isolate individuals, lead to over-abundance of fauna in a restricted habitat and result in in-breeding (Jaeger and Fahrig 2004; Muir 2008; Balkenhol and Waits 2009). It has been suggested that increased levels of predation could potentially be caused by the funnelling effect of fauna exclusion fences along roads to underpasses (Fahrig *et al.* 1995). Many of these impacts can ultimately lead to impacts on species at a genetic level.

Jaeger and Fahrig (2004) studied the effects of road fencing on populations and recommended that fauna exclusion fences not be used along roads if the target fauna population is likely to remain stable or increase. Fauna exclusion fences are discouraged along roads where species need access to habitat resources on both sides of the road, unless land bridges or underpasses are incorporated.

Fauna exclusion fences can result in fauna becoming trapped within the road reserve. Escape routes, such as poles against the fence, should be installed to allow trapped fauna to escape, especially in bushfire-prone areas. Animals becoming entangled in the fence may also be a problem (e.g. cassowaries in northern Queensland) (Chapter 8).

Vegetation adjacent to fauna exclusion fencing needs to be removed or otherwise maintained to ensure effectiveness. This can be undesirable if the fence is proposed to be erected within sensitive vegetation or habitat areas. Surface water flow can also be disrupted by the placement of the fauna exclusion fencing, leading to changes in vegetation or new opportunities for environmental weed species to flourish.

Coverings

Overview	
What:	Coverings can physically prevent many types of fauna from entering an area.
Where:	Useful where development (e.g. aquaculture, mining or airports) contains a localised hazard (or attractant) that can be covered.
Pros:	Protects fauna from the dangers of a localised hazard or reduces fauna presence in a hazardous area.
Cons:	Possible entanglement, although this can be minimised by careful selection of material and correct installation.
Cost:	Low–high, depending on the area to be covered.
Companion measures:	Habitat surrounding the development can be made less attractive to discourage animals (this chapter).
Alternative measures:	Avoiding an impact by removing the hazard (Chapter 4); visual deterrents and auditory deterrents (this chapter).

Description and potential applications

Coverings (e.g. netting and cages) are physical structures that can prevent birds and other fauna from accessing a component of a development. Coverings can be physically applied to a localised hazard (Figure 7.7). Alternatively, coverings can be used to reduce access to attractive habitat (e.g. food, water and shelter) to discourage fauna visitation to the development site. Reducing attractive habitat is described further later in this chapter.

Coverings provide a high level of confidence that fauna will not be able to access a hazard or attractant. Coverings have been used in various ways by development, including:

- to prevent fauna accessing hazardous material in pond structures within mining developments (Donato *et al.* 2007)
- to reduce the potential for bird mortality from collision (e.g. netting in front of windows that represent high risk; Pfennigwerth 2008)
- to cover holes that can act as 'pitfall traps' for ground dwelling fauna (Lloyd *et al.* 2002; Pedler 2010)
- to reduce attractive habitat near threats, such as netting that is commonly used to prohibit birds accessing water bodies near airports (Australian Transport Safety Bureau 2010).

Common types of coverings include: netting, caging, overhead wires and high-density polyethylene balls (Table 7.3). Other coverings applied to small areas include plastic sheeting and shade cloth.

Figure 7.7. Water drains, such as these at Newcastle Airport, New South Wales, can be caged (Photo: J Gleeson)

Netting is the most widely used covering because it can be used in a range of applications. Thin nylon (monofilament) should be avoided because animals can become entangled, injured or killed (Nemtzov and Olsvig-Whittaker 2003). Overhead wires are a cheaper alternative to netting and are useful in situations where minimising, but not excluding, birds is required (Belant and Ickes 1996). Some birds tend to quickly learn how to avoid the wires and still access the habitat (Barlow and Bock 1984). Caging and polyethylene balls are more expensive, but both options have unique advantages. Caging is strong and requires little maintenance, while polyethylene balls provide a solid cover over liquid, reducing unwanted algal growth.

Coverings are widely used throughout Australia and around the world. There are various companies in Australia that specialise in the supply and installation of coverings.

Table 7.3. Common types of coverings

Type of covering	Description
Netting	Typically made from high tensile UV-treated polyethylene. Mesh size and grade can vary depending on the animal to be excluded.
Caging	Best when made from stainless steel (non-rusting) wire and mainly used at ground level.
Overhead wires	Consist of a network of stainless steel wires suspended horizontally in one direction or arranged in a grid pattern.
Polyethylene balls	Air-filled plastic balls that float on the surface of a liquid. Closely packed on the top of a pond they prevent birds from accessing the contents of the pond.

Effectiveness

When designed and applied correctly, coverings can be a reliable and humane way to exclude fauna (Bomford and Sinclair 2002). Factors likely to influence the effectiveness of coverings include its design, placement and maintenance. These factors are discussed below.

Design

Entanglement of fauna, mainly birds, can occur when netting or overhead wires are not correctly designed and applied. To reduce the chances of birds becoming entangled, netting should be tensioned to remove slack (Donato *et al.* 2007), the mesh should be small (e.g. 5–7 cm; Nemtzov and Olsvig-Whittaker 2003) and the netting should be maintained to ensure that any tears are repaired quickly. Overhead wires can entangle and harm birds, particularly in low light due to lower visibility. Unfortunately, as the thickness of wire increases, it is also more likely that birds will use the strands for roosting. Overhead wires should be made from a heavy duty wire to avoid breakages (Marsh *et al.* 1991).

The shape and size of the area to be covered will dictate which sorts of coverings can be used. This is particularly true for water bodies. Netting over water requires a system of tensioned wire rope so that the netting does not sag into the water. A rectangular pond structure is easier to cover than a square pond covering the same area due to the ability to tension the narrower width (Martin and Hagar 1990). Netting, caging or using overhead wires over larger water bodies is technically difficult to construct and maintain based on current technologies. In these situations, different measures may be suitable to deter fauna, such as fauna exclusion fencing and reducing attractive habitat in combination with sensory measures (this chapter).

If polyethylene balls are to be used on a pond structure, a sufficient number of balls will be required to cover the surface to effectively exclude birds. Donato *et al.* (2007) conducted a review of the effects of tailings solutions on animals and found that high-density polyethylene balls are effective, but can become bogged when used on more dense liquids, such as slurry.

Placement

The effectiveness of coverings can be dependent on placement. For example, the effectiveness of overhead wires at excluding birds is partly dependent on whether the area below the wires is desirable habitat (Marsh *et al.* 1991). The more desirable the habitat, the more persistent the birds are likely to be at trying to access the area. This is also likely to apply to netting if the animals are able to squeeze through the mesh.

Maintenance

Netting and overhead wires are susceptible to breakage and will require frequent inspections and maintenance. It is important that these are maintained so that animals are not able to break through and become trapped underneath the covering. Reducing attractive habitat (discussed later in this chapter) is a companion measure that can minimise fauna interacting with the coverings, thus reducing maintenance requirements.

The effectiveness of netting, overhead wires and polyethylene balls can be drastically affected by extreme environmental conditions (e.g. flooding or strong wind). Stainless steel cages are, of course, strong and likely to be more effective than other types of coverings in these types of situations. If other types of coverings are used in areas affected by extreme environmental conditions, they should be inspected frequently and repaired quickly when damaged.

Impacts on other aspects of the environment

Applied correctly, coverings have minimal impacts on other aspects of the environment. Possible death or injury from entanglement in netting or overhead wires is perhaps the greatest potential impact. However, the potential for this impact can be minimised by using the correct materials and installing the coverings correctly (e.g. by removing slack).

If coverings are proposed to be used to exclude animals from their habitat resources, the impact of barring the resources should be considered. Barring habitat resources will likely cause fauna to move to find alternative resources, if available.

Reducing the attractiveness of fauna habitat

Overview	
What:	Removing attractive habitat to deter fauna from occurring near a hazard.
Where:	Useful where development (e.g. roads, mining or airports) poses a hazard to fauna that are attracted to the development site.
Pros:	May prevent injuries or fatalities of animals.
Cons:	Requires simplification or removal of habitat that can adversely impact native wildlife.
Cost:	Medium.
Companion measures:	Fauna exclusion fencing, visual deterrents and auditory deterrents (this chapter)
Alternative measures:	Avoiding an impact by removing the hazard (Chapter 4); fauna exclusion fencing and sensory deterrents (this chapter).

Description and potential applications

Removal of habitat or modifying it so that it is less attractive to fauna can be a way to deter fauna from a development. This may seem counterintuitive given that wildlife conservation usually involves protection and enhancement of land for habitat. However, keeping animals away from the hazardous components of developments by removing habitat and maintaining less attractive habitat can be a legitimate way to conserve some fauna. An example of this is land surrounding air strips that is kept as mown grass so that seed eating birds are not attracted (Figure 7.8; Box 7.1). This is done to reduce the number of collisions between aircraft and birds.

Figure 7.8. Mown grass at the Adelaide Airport, South Australia (Photo: J Gleeson)

Reducing attractive habitat may be necessary when fauna are at risk from trying to access habitat resources in the development site (e.g. birds attempting to access water in a pond containing hazardous material). Attractive habitat outside of a development site may also need to be managed in some way to reduce impacts on wildlife. For example, Blackwell *et al.* (2009) stressed the importance of thoughtful land use planning around airports to reduce fauna collisions with aircraft (e.g. management of stormwater to avoid open water that attracts birds) (Box 7.1).

The first step in reducing attractiveness of habitat is to identify what is drawing the fauna to the area. It is likely that the attractant falls into one of three main categories: food, water or shelter. Possible food sources include vegetation, fruit, seeds, invertebrates or other vertebrate fauna. Open water can attract a range of fauna, particularly waterfowl and shorebirds. For example, Leach (1994) found that the number and diversity of birds using farm dams in Queensland increased with the area of open water. Shelter is used for roosting/perching, predator avoidance and breeding. Possible sources of shelter include vegetation, rock piles and roosting structures. Some attractants can vary temporally, often due to season or climatic conditions.

Measures to reduce attractiveness of habitat are widely used throughout Australia and around the world. Some examples are provided in Table 7.4. It is relevant to note that this table includes only a few examples to give a general idea of different ways to reduce the attractiveness

---

Box 7.1. Management of habitat to reduce aircraft–bird collisions

Collision between aircraft and fauna (mainly birds and bats) continues to be an important safety issue for the aviation industry around the world, with most bird strikes involving aircraft occurring at, or near, airports (Australian Transport Safety Bureau 2003, 2010). Collisions between fauna and aircraft are rarely discussed in the media, so it is surprising that between 2002 and 2009 there were 9287 strikes reported in Australia alone (Australian Transport Safety Bureau 2010). The most commonly struck animals were lapwings and plovers (597), bats and flying foxes (542), galahs (532), kites (453) and magpies (353) (Australian Transport Safety Bureau 2010). These statistics are concerning, considering that aircraft movements are increasing globally and anticipated to be around 120 million annually by 2025 (Aaronson 2007).

At each airport, there is a requirement to keep fauna away from the runway, aircraft approach and departure paths, as well as adjacent areas. The risk of aircraft collision with birds and bats varies between airports due to differences in operations, weather conditions, the species present and their abundance (Steele 2001), so solutions to reduce collisions must be tailored to suit each situation.

The aviation industry uses various measures to deter fauna from airports (e.g. auditory deterrents such as cracker shot and live shotgun rounds; Australian Transport Safety Bureau 2003), but reducing attractive habitat at airports is the most common and effective strategy used to tackle the issue (Transport Canada 2002; Australian Transport Safety Bureau 2010).

Specific measures to reduce attractive habitat used in Australia include minimising nesting areas, reducing open water, maintaining grass length and minimising foraging resources. Using a combination of all of these measures has been proved to reduce risks, but not entirely eliminate them (Australian Transport Safety Bureau 2010).

Table 7.4. Examples of ways to reduce attractiveness of habitat

Measure	Description
Manipulation of vegetation	Removal or modification of vegetation to make it less attractive to fauna.
Invertebrate control	Reduction of invertebrates that provide food for other fauna.
Removal of carrion	Collection and disposal of the carcasses of dead animals (e.g. road-kill) so they do not attract scavengers.
Management of open water	Removal or reduction of open water that may attract fauna (e.g. waterbirds).

of habitat. Indeed, there are many effective ways to reduce the attractiveness of habitat, because the most appropriate method depends on the species and the characteristics of the habitat present at each particular development site.

By providing alternative habitat in a nearby location, such as decoy wetlands (Read 1999), foraging areas (Martin 2011), decoy crops or revegetated areas (Bomford and Sinclair 2002), it could be possible to divert fauna away from a development site. A second benefit of doing this is that fauna deterred from a development site are re-accommodated elsewhere. Animals are more likely to persist in using habitat in a development site if habitat resources are limited nearby (e.g. open water at a development site is more attractive to fauna when the landscape is experiencing drought conditions).

It is not only natural habitats that are attractive to fauna. For example, open buildings at development sites often attract introduced birds (e.g. rock dove *Columba livia*, common starlings *Sturnus vulgaris* and house sparrows *Passer domesticus*) which can compete with native species. Similarly, poor waste management at a development site can attract certain types of fauna, leading to increases in populations of common species. For example, increases in birds attracted to rotting garbage in landfills (e.g. Australian white ibis *Threskiornis molucca*) (Smith and Carlie 1993; Cook *et al.* 2008).

Manipulation of vegetation

Vegetated areas can be made less attractive to fauna by:

- simplifying habitat by removing select food trees
- maintaining long grass (e.g. >15 cm) to interfere with visibility, feeding ability and/or movement (Brough and Bridgman 1980; Barras and Seamans 2002) or, in other situations, maintaining short grass (e.g. <15 cm) to interfere with predator avoidance and reduce the availability of food, such as seed (Figure 7.9), and shelter (Smith and Carlile 1993)
- replacing attractive vegetation with less attractive vegetation
- removing all vegetation from an area.

Clear zones (i.e. areas with no vegetation) can be maintained along the sides of roads to reduce the potential for vehicle collision with fauna. Grassland birds, for example, that may be flushed into oncoming traffic would be at less risk if grass was kept short by the sides of the road. Roadside vegetation can be altered to make it less attractive for foraging kangaroos to reduce kangaroo–vehicle collisions on outback highways (Klocker *et al.* 2006).

Vegetation is aesthetically pleasing, so replacing existing vegetation with alternate vegetation that is less attractive to animals (e.g. certain non-native varieties) can be a more desirable way of reducing attractive habitat than removing vegetation altogether. Further, replacing vegetation with paving or concrete may be counter-productive because the thermal heating of these surfaces can attract basking animals (e.g. reptiles).

Figure 7.9. Galah feeding on flowering gomphrena weed (*Gomphrena celosioides*) (Photo: D Gleeson)

A grass and fungus combination (endophytic fungal symbionts of grasses) that is less palatable to birds and insects and has a low seed production is a promising new way to reduce attractiveness of grassed areas (Australian Transport Safety Bureau 2008; Pennell *et al.* 2010). This grass is being trialled in New Zealand at Christchurch International Airport and has been successful in reducing geese and other birds (Pincock 2006).

Invertebrate control

Controlling invertebrates can remove food sources for birds. Soil invertebrates, such as earthworms have been controlled using chemical treatments (Brough and Bridgman 1980; Transport Canada 2002). The possible adverse impacts on invertebrate species would need to be weighed against the need to reduce food sources for birds. For other applications, the use of orange lights attracts fewer insects (e.g. moths) than white lights (van Tets *et al.* 1969).

Removal of carrion

Carrion (the carcass of a dead animal) is food for scavengers such as birds of prey. For example, road-kill attracts scavengers and, while they are feeding on the road, these scavengers are at risk of collision with motor vehicles (Huijser *et al.* 2008). To prevent death of scavengers, regular inspections would be required and any road-kill found collected and disposed of. In the United States of America, deer carcasses on the side of the road are covered with composting material (e.g. compost, wood chip or sawdust) to allow the road-kill to decompose naturally (Montana Department of Transportation 2007).

Management of open water

Open water is present in dams, drains, ponds or ditches constructed for site water management or water-holding ponds for use in the development (e.g. dust suppression). Open water at a development site can be managed to make it less attractive to fauna by:

- designing an efficient site drainage with the aim of avoiding creation of open water ponds

- allowing water to quickly drain into covered pond structures
- designing necessary pond structures to have minimal area of open water and banks graded to hinder birds landing.

In addition to the above, Donato *et al.* (2007) reported that reducing the supernatant surface area in a tailings storage facility is a successful way to reduce fauna interacting with potentially hazardous tailings.

Effectiveness

Reducing the attractiveness of habitat is often a primary measure to deter fauna due to its effectiveness. This depends on how species use the habitat and, the availability of alternative habitat nearby. The effectiveness of other deterrent measures (e.g. visual deterrents) can also be enhanced by reducing attractive habitat (Donato 1999; Donato *et al.* 2007).

Removal of trees and maintenance of a particular grass length can be effective, but the optimal grass length is species-dependent. From a study of 13 airfields in the United Kingdom, Brough and Bridgman (1980) found that keeping grass long effectively reduced bird numbers, including lapwings, woodpigeons, starlings and gulls, whereas Smith and Carlile (1993) studied control methods for the silver gull (*Larus novaehollandiae*) and found mowing long grass at silver gull colony sites before the commencement of breeding decreased both the density of nests and nesting success. Maintaining long grass and maintaining short grass are both used to reduce attractive habitat at Australian airfields, with 64% of all aerodromes maintaining long grass (Australian Transport Safety Bureau 2010).

The effectiveness of reducing attractive habitat to avoid birds colliding with stationary structures is questionable. Klem *et al.* (2009) found that removing vegetative ground cover adjacent to buildings did not significantly reduce the risk of birds colliding with the buildings.

Impacts on other aspects of the environment

If native vegetation is to be modified to reduce its attractiveness to fauna, the resulting impact of loss of habitat on wildlife must be evaluated. Reducing the attractiveness of naturally occurring habitat can have a range of implications for ecosystem processes, species composition, life cycles, populations and movement. For example, it may be possible to reduce the number of animals killed by motorists by maintaining clear zones along the sides of roads. However, some species that would have crossed the road before vegetation removal will be unwilling to cross such an open area. Ultimately this fragmentation of habitat could cause a population to become unviable: a far greater impact than the occasional vehicle strike fatality. Hence, in many local situations, reducing attractive habitat will be unacceptable and alternatives will need to be investigated, such as fauna exclusion fencing, covering the hazardous component, visual deterrents and auditory deterrents.

Studies overseas suggest that maintaining long grass can result in an increase in rodents and potentially an increase in raptors (Brough and Bridgman 1980; Transport Canada 2002). Exotic animal prevention and control may reduce the potential for this to occur. Obviously, any chemical control should be restricted to avoid adverse impacts on the environment. This especially applies to chemical applications to control invertebrates to reduce impacts on birds.

It is plausible that removing large carrion may be counter-productive in reducing road-kill if carrion can act as a warning for drivers to slow down for fauna. However, leaving large road-kill on the sides of roads is unsafe for road users and the presence of road-kill should not be depended upon to reduce the incidence of vehicle–fauna collision. Various other measures in this book are aimed at reducing road-kill, such as modifying human behaviour (Chapter 5) and fauna exclusion fencing (discussed earlier in this chapter).

Visual deterrents

Overview	
What:	Visual stimuli intended to deter fauna (mainly birds) from a hazardous component of a development.
Where:	Used where development (e.g. powerlines) poses a hazard to fauna.
Pros:	Those designed to aid navigation around obstacles can prevent injuries or fatalities. Those that are designed to scare fauna may be effective initially.
Cons:	Many types are poor at deterring fauna over an extended period of time.
Cost:	Low.
Companion measures:	Reducing the attractiveness of fauna habitat, chemical deterrents and auditory deterrents (all this chapter).
Alternative measures:	Avoiding an impact by removing the hazard (Chapter 4); fauna exclusion fences and coverings (this chapter).

Description and application

Visual deterrents are visual stimuli intended to deter fauna, particularly birds. A scarecrow was perhaps the first visual deterrent – used for centuries to scare birds away from agricultural crops. Fauna use visual stimuli to navigate, detect habitat resources, socially interact and avoid perceived threats (e.g. predators). There are two broad groups of visual deterrents: those that are designed to scare fauna to move on to another location and those that aid the navigation of animals (usually birds) around obstacles to reduce the risk of collision (e.g. bird flight diverters installed on overhead cables). Visual deterrents are generally brightly coloured, reflective and/ or large in an attempt to improve visibility of a hazard to fauna.

Visual deterrents are common because they are inexpensive and can require little effort to install and maintain. Visual deterrents are widely used in Australia across various types of infrastructure (e.g. roads, tailings dams, sewerage ponds, powerlines and on buildings).

There are countless types of visual deterrents available on the market and it can be difficult to choose the best for a given situation, particularly because many have not been scientifically tested. Examples of visual deterrents are described in Table 7.5 and further below. In addition to those listed, flags, streamers and reflective objects have been used to deter fauna. Other visual deterrents not listed may be worth trialling and there is always room for further ingenuity.

Bird flight diverters

Bird flight diverters are attached at intervals along powerline conductors with the intention of making overhead lines more visible to birds to avoid collision (Box 7.2). Types of marking devices include aerial warning spheres, spiral bird flight diverters, swinging plates and flappers (IEEE Task Force on Reducing Bird Related Power Outages 2004).

Aerial warning spheres are generally not used for reducing bird collision because they are one of the least cost-effective options and the weight of these diverters will limit the frequency with which they can be used along a powerline. Aerial warning spheres are balls (~30 cm or 60 cm in diameter) primarily used to warn humans of the presence of powerlines (e.g. crop dusters in agricultural areas). These spheres are usually bright orange, but other colours are also used.

Spiral bird flight diverters are an economical option and are increasingly preferred by Australian electricity companies. Spirals are typically made from a solid polyvinyl chloride (PVC) material in a cone shaped spiral (Harness and Carlton 2001). The spiral is twisted around the conductor and remains fixed in the applied position once it is installed on the line.

Table 7.5.   Examples of visual deterrents

Type of visual deterrent	Description
Bird flight diverters	Objects that are attached to an obstacle (e.g. powerlines) to improve visibility of the obstacle so that birds are less likely to collide with it.
Flashing lights, strobe lights and lasers	Lights used to repel nocturnal and diurnal animals.
Predator/human effigies	A life-sized effigy of a predator of the target species or a human effigy have been used to frighten fauna.
Window coverings	Window coverings have been designed to prevent birds from flying into windows.
Disturbance	Chasing animals away from a potential hazard with a disturbance (e.g. vehicle).
Warning reflectors	Reflectors mounted on posts at regular intervals along a road to reflect the headlights of oncoming vehicles into the eyes of animals on the road or road verge have been used with the intent of preventing fauna–vehicle collisions.

Swinging plates and flappers are free to pivot on the fixing. The motion component allows wind to pass through, reducing resistance on the powerline. Some are also designed to emit ultraviolet light.

Flashing lights, strobe lights and lasers

Lights have been used with the intention of repelling nocturnal animals and diurnal animals. Technology involving automated lasers and hand-held lasers has been developing over the last 20 years (e.g. Briot 2005; Australian Transport Safety Bureau 2010). However, there is a perceived sensitivity in using lasers around aircraft (Australian Transport Safety Bureau 2008).

Light purposefully directed with the aim of deterring waterfowl from water bodies has also been used. A rotating, intermittent beacon directed at a shallow angle over the surface of a water body has been used successfully at the Roxby Downs sewerage ponds in arid South Australia (Read 1999).

Predator effigies

Stationary or mechanised effigies of predators are sometimes used in an attempt to simulate a species' response to the threat of predation. Predator effigies are sometimes used in conjunction with the broadcasted calls of the predator, which is intended to act as an auditory deterrent (Ronconi and St. Clair 2006).

Live animals (i.e. hawks) have also been used to deter prey, not always resulting in prey fatality (e.g. to chase birds from landfills; Baxter and Allen 2006). Using live raptors has been recommended overseas as a way to reinforce visual deterrents (Transport Canada 1998; Cook *et al.* 2008). Cook *et al.* (2008) found that different species of raptor can be more or less effective as deterrents, depending on the target species.

Window coverings

UV-reflecting and UV-absorbing window coverings are designed to prevent birds from flying into windows (Klem 2009). The films allow one-way viewing, so do not obstruct the view from within the building. The windows appear opaque from the outside and so birds adjust their flight behaviour. Window markings (such as vertical dark stripes) have also been used with the intention to reduce bird collisions with glass (Iuell *et al.* 2003).

Box 7.2.   Powerline design to reduce risk of bird collision

Electricity is integral to modern society. It is distributed though an estimated 65 million kilometres of medium–high voltage powerlines throughout the world (ABS Energy Research 2008). Bird interactions with powerlines can cause injuries (e.g. wing breakage) and death (McNeil *et al.* 1985; Rusz *et al.* 1986; Alonso *et al.* 1994; Janss 2000; Rubolini *et al.* 2005) and also power outages. The factors that that lead to birds colliding with powerlines are temporally and spatially variable (Martin and Shaw 2010). Environmental factors such as weather conditions, season and time of day influence risk of collision (Bevanger 1994; IEEE Task Force on Reducing Bird Related Power Outages 2004).

Some recognised characteristics that make birds more susceptible to collision include migrating at night (Rubolini *et al.* 2005), large size and wide wingspan (Bevanger 1998), poor manoeuvrability (Janss 2000), low flight height (Rollan *et al.* 2010) and restricted field of vision during flight (Martin and Shaw 2010; Martin 2011). Birds that display some of these characteristics and are typically victims of collision include the Australian pelican (*Pelecanus conspicillatus*) (Figure 7.10), cranes, storks, herons, egrets, ibises and spoonbills (Bevanger 1998). Recent advances in radio-tracking and programs to model movement of species have been used to model projected risk of collision with powerlines in South Africa to help reduce the impact of powerlines on birds (Shaw *et al.* 2010; Rollan *et al.* 2010).

Figure 7.10.   An Australian pelican (*Pelecanus conspicillatus*). Its large size and wide wing span make it susceptible to collision with powerlines (Photo: R Whitney)

Bird flight diverters are viewed as one way to reduce bird collisions with powerlines; however, the effectiveness of bird flight diverters has been reported to be variable. During a recent study of bustards, cranes and storks, Martin and Shaw (2010) found that these birds have a limited visual field when flying. This renders them blind in the direction of travel, preventing powerlines or bird flight diverters from being seen. Notwithstanding, Martin (2010) described increasing the conspicuous nature of hazards as valuable. Visual deterrents should be in high contrast to the hazard, incorporate movement and be large for greater visibility (Martin 2010).

Jenkins *et al.* (2010) undertook a review of 13 studies on bird flight diverters and concluded that generally any sufficiently large bird flight diverter (which thickens the appearance of the line by at least 20 cm over a length of at least 10–20 cm), placed at 5–10 m is likely to lower general collision rates by 50–80%. There is, however, a lack of experimental studies on the medium and long-term effectiveness of particular bird flight diverter designs (Jenkins *et al.* 2010). In any case, their effectiveness would be situation-dependent and they are likely to be more useful when used in flight paths and near key habitat areas.

Reflective properties may enhance the visibility of bird flight deterrents. Increasing the visibility of the deterrents at dusk and during the night might increase their effectiveness in deterring some species (e.g. waders) (Boag and Lewin 1980).

Good powerline placement and design is fundamental when attempting to reduce bird collisions because visual deterrents do not remove all risk. Important design factors include:

- installation of powerlines underground (McNeil *et al.* 1985; Jenkins *et al.* 2010)
- avoiding installation of powerlines near key habitat areas or open bodies of water (e.g. minimising creek crossings) (McNeil *et al.* 1985; Harness and Carlton 2001; Rollan *et al.* 2010)
- using topography and natural barriers (e.g. vegetated areas) to avoid flight paths;
- aligning powerlines parallel to, rather than across, flight paths (e.g. prevailing winds for migratory species) (McNeil *et al.* 1985)
- aligning the conductors on one horizontal plane (Harness and Carlton 2001).

Removal of the earth wire has been suggested by some authors to reduce collision risk, but this increases the risk of fire and is therefore not a viable option in many cases (Harness and Carlton 2001; Jenkins *et al.* 2010).

Disturbance

Chasing animals away from a potential hazard has been used as a method of deterring them from an area. For example, driving along runways before aircraft land has been done to flush kangaroos from the path of the aircraft. Another example, is using radio-controlled model airplanes to chase birds from hazards (Loud 2000). Physically disturbing animals generally has only a short-term application because it relies on an operator to chase the animal continually.

Warning reflectors

Animal warning reflectors are intended to reduce the risk of nighttime collisions between vehicles and fauna. Reflectors are mounted on posts at regular intervals along the road verge and reflect vehicle headlights to create a low-intensity red beam that acts as a moving 'lighted fence' (Forman 2003). Warning reflectors have been around since the early 1970s and are usually red, but do come in other colours. They are similar to the reflectors used to alert motorists to the edge of the road.

Effectiveness

Various authors (e.g. Martin and Hagar 1990; Ronconi and St Clair 2006; Donato *et al.* 2007) have reported that visual deterrents used to scare fauna work initially, but are not very effective in the long term because target species often readily habituate to a particular deterrent. Conversely, visual deterrents used to aid the navigation of birds around obstacles have shown to be effective in some situations, such as bird flight diverters, as discussed in Box 7.2, and UV-reflecting and UV-absorbing window coverings (Klem 2009). There is, however, a lack of robust scientific experiments on visual deterrents used to aid the navigation and deterrence of Australian birds. For this reason, visual deterrents should be used on a trial basis.

Because there are many different variations of visual deterrents available, and they are all used in different situations (i.e. different environments to deter different species), the success or failure of specific devices cannot be generalised to all visual deterrents. It is clear, however, that visual deterrents alone should not be relied upon to deter fauna from a hazard at a development site. One of the biggest reasons visual deterrents used to scare fauna tend to have limited efficacy is due to habituation (Boag and Lewin 1980; Bomford and O'Brien 1990; Ujvari *et al.* 1998), which is discussed further in Box 7.3.

The effectiveness of bird flight diverters would be situation-dependent; they are likely to be more useful when used in flight paths and near key habitat areas (Box 7.2). It is little surprise that the use of stationary raptor models to scare birds from powerlines by Janss *et al.* (1999) was ineffective because of habituation. Before applying any deterrent technology, it is wise to seek evidence from the supplier that the deterrent has been effective elsewhere in a similar circumstance to where it is proposed to be used.

Box 7.3.   Overcoming habituation of sensory deterrents

Although it is generally positive for animals to become accustomed to nearby development, it may become an issue when it is desirable to deter fauna from a development site for their safety. Visual and acoustic deterrents used to scare fauna often have limited efficacy because animals can quickly become accustomed to and ignore (i.e. habituate to) the stimulus (Boag and Lewin 1980; Bomford and O'Brien 1990; Ujvari *et al.* 1998). Species in some situations are less likely to habituate than others. For example, nomadic species in remote areas are less likely to habituate to a deterrent because they are more likely to move on to a different area (Read 1999).

The following measures may lessen habituation:

- using an 'integrated hazing system' with multiple types of deterrents (e.g. combination of visual and auditory deterrents)
- producing a variety of sounds in random or pre-programmed order at random timed intervals (International Bird Strike Committee 2006). Visual and acoustic deterrents are less likely to be effective if they are predictable to the target species.
- deploying deterrents in response to specific animal activity
- using portable auditory deterrents (handheld or vehicle mounted), which are generally regarded as more efficient than stationary auditory deterrents (International Bird Strike Committee 2006; Australian Transport Safety Bureau 2008).

Some authors have also suggested reinforcing deterrents with real danger (usually shooting) (e.g. Bomford and Sinclair 2002), although anecdotally this has not been found to be particularly effective in Australia.

Ramp and Croft (2006) found that the behavioural response of captive Australian macropod species to the reflectors on a simulated road was negligible and concluded that using reflectors to reduce collisions with kangaroos and wallabies would be of little value. Published studies from other countries have been mainly for deer and many have found that reflectors are not very effective (Waring *et al.* 1991; Reeve and Anderson 1993; Ujvari *et al.* 1998; D'Angelo *et al.* 2006).

The nature of the immediate behavioural reactions to a visual deterrent is often species-specific. An understanding of the behaviour of the species that are to be targeted is important when deciding if a visual deterrent is suitable for an application. For example, a rotating, intermittent beacon directed at a shallow angle over the surface of the Roxby Downs sewerage ponds in arid South Australia was effective in deterring most species of waterfowl and appeared to decrease waterfowl abundance by more than 90% (Read 1999). However, the reaction to the deterrent of both the hoary-headed grebe (*Poliocephalus poliocephalus*) and Australasian grebe (*Tachybaptus novaehollandiae*) was to dive under the water rather than to fly away (Read 1999). This deterrent may not be suitable for these particular species when applied to toxic water bodies such as mine tailings dams. Further to this, moulting or unhealthy waterfowl in general were not deterred and nor were certain species of wading birds, such as masked lapwings (*Vanellus miles*), black-fronted dotterels (*Elseyornis melanops*) and red-necked avocets (*Recurvirostra novaehollandiae*) (Read 1999).

Impacts on other aspects of the environment

Visual deterrents intended to deter fauna from a hazardous component of a development site are not likely to impact species present outside of the development.

Chemical deterrents

Overview	
What:	Chemical stimuli intended to deter fauna from occurring near a hazardous component of a development.
Where:	Useful where development (e.g. roads) poses a hazard to fauna.
Pros:	May be tailored to deter specific fauna to prevent injuries or fatalities.
Cons:	Ethical issues due to the potential side-effects on wildlife.
Cost:	Low to high, depending on development size.
Companion measures:	Reducing the attractiveness of fauna habitat, visual deterrents and auditory deterrents (all this chapter).
Alternative measures:	Avoiding an impact by removing the hazard (Chapter 4); fauna exclusion fences and coverings (this chapter).

Description and application

Chemical deterrents are chemical compounds intended to deter fauna from occurring near a hazardous component of a development. Depending on the type, chemical deterrents are either ingested, smelled or touched by fauna and, because they are unpleasant, fauna tend to leave the affected area. Fauna naturally use smell and taste to detect habitat resources, socially interact and avoid perceived threats (e.g. predators).

Chemical deterrents have mainly been used in agriculture to deter vertebrate and invertebrate pest species (Bramley and Waas 2001; Bomford and Sinclair 2002). They are not widely used on native species in Australia to reduce the impact that development may have on them.

Table 7.6.   Examples of chemical deterrents

Type of chemical deterrent	Description
Olfactory deterrents	Chemicals that produce predator odours (real or synthetic) to scare fauna or chemicals that irritate fauna so they move away from the source of the irritant.
Taste deterrents	Chemicals are applied to a resource, such as food or water, to make it unpalatable.

However, there is interest from industry in Australia to improve ways of deterring fauna from hazardous components of developments and chemical deterrents are one possible method that is currently under-studied (Queensland Department of Transport and Main Roads 2010).

Chemical deterrents here are split into two broad categories: olfactory deterrents and taste deterrents (Table 7.6). A third type of chemical deterrent, 'tactile deterrent', is also used to discourage fauna, but more so for human purposes rather than for the welfare of the animal. For example, commercially available tactile deterrents are applied to structures to discourage birds from perching on them.

Effectiveness

Chemical deterrents are not generally used in Australia to reduce impacts of development on fauna. One reason chemical deterrents are not widely used is the ethical issue of using chemicals to change fauna behaviour. The other reason is the lack of research supporting the effectiveness of chemical deterrents.

The effectiveness of a chemical deterrent will depend on how it is applied, the receiving environment and the fauna targeted by the repellent. Various authors in Australia have reported that chemical deterrents tend to be less effective in the natural environment (Bomford and Sinclair 2002; Bishop *et al.* 2003). This may be due to competing sensory stimulation outside laboratory conditions.

It has been shown that even similar species do not always respond to the same chemical deterrent in the same way. For example Ramp *et al.* (2005b) studied the use of a synthetic predator scent that mimics dog urine and found that, while wallabies fled from the scent, pademelons approached the scent to investigate it.

Stevens *et al.* (2000) tested chemicals that irritate fauna as part of a successful integrated hazing system that included pyrotechnics and acoustic deterrents. Unfortunately, due to the experimental design, they could not attribute the effectiveness of the system to any one type of deterrent. Notwithstanding, during pilot trials, waterfowl reacted to the repellent cloud by moving away from the source and plume (Stevens *et al.* 2000). Chemical repellents are generally expensive and their application is often labour-intensive (Stevens *et al.* 2000; Bishop *et al.* 2003).

Overseas studies have reported that animals can quickly learn to identify and avoid chemical laden baits (Bishop *et al.* 2003) or become desensitised (Conover 2001). Short-term use may help overcome habituation to chemical deterrents.

Impacts on other aspects of the environment

Some chemicals can deter fauna but also adversely affect them. For example, some smells can cause pregnant animals to abort their foetus. This, of course, could have consequences for species abundance and their populations. Only those chemicals registered as chemical deterrents in Australia should be used here, and only as specified. It is not known whether chemical deterrents derived from natural sources, such as predator smells, are less likely to have an impact on other aspects of the environment.

Auditory deterrents

Overview	
What:	Auditory stimuli (i.e. sound) intended to deter fauna from a hazardous component of a development.
Where:	Useful where development (e.g. roads, airstrips and tailings dams) poses a hazard to fauna.
Pros:	May improve the effectiveness of other deterrents.
Cons:	Noise may be a nuisance, especially near residential areas.
Cost:	Low.
Companion measures:	Reducing the attractiveness of fauna habitat, visual deterrents and chemical deterrents (all this chapter).
Alternative measures:	Avoiding an impact by removing the hazard (Chapter 4); fauna exclusion fences and coverings (this chapter).

Description and application

Auditory deterrents produce sounds intended to scare fauna away from a hazardous component of development. In nature, animals use auditory stimuli to socially interact and to avoid perceived threats (e.g. predators). Some auditory deterrents are designed around biologically significant sounds (e.g. distress and predator calls), while others emit an artificial sound that has no social context.

Auditory deterrents are widely used in Australia. They are used at developments such as roads, airstrips and tailings dams. Auditory deterrents, along with visual deterrents, are often used for managing birds (Berge *et al.* 2007; Soldatini *et al.* 2008). In fact, various studies have found auditory deterrents to be more effective than visual deterrents in comparable situations (e.g. Cook *et al.* 2008).

There are numerous types of auditory deterrents available on the market. It is difficult for a consumer to choose which would be optimal for a given situation, particularly as many of them are not scientifically tested on Australian fauna. For example, Bomford (1990) tested a particular brand of sonic device reported to be effective on birds in Australia and they found that it was ineffective during the experiment.

Common types of commercially available auditory deterrents are described in Table 7.7 and below. Auditory deterrents are either portable (e.g. hand-held distress and alarm calls) or static (e.g. broadcasting of distress and alarm calls from radio-controlled sound generator).

Distress and alarm calls

When animals are stressed or threatened, they may emit a distress or alarm call. Recorded distress and alarm calls are broadcast to frighten and repel fauna, usually individuals of the same species. Radio-controlled sound generators enable an operator to activate the calls when fauna approach a hazard.

Predator calls

The sound of a predator can cause prey to flee in order to avoid a perceived threat. Recorded predator calls are either used on their own or sometimes in conjunction with an effigy of the predator to deter fauna (Ronconi and St. Clair 2006). The use of predator calls to deter a particular species requires intimate knowledge of a species' behaviour in response to predator cues, because species react in different ways to predator calls (e.g. animals may freeze or run).

Table 7.7.   Common types of auditory deterrents

Type of auditory deterrent	Description
Distress and alarm calls	Broadcasting recorded distress and alarm calls of particular species to frighten and repel fauna, usually individuals of the same species.
Predator calls	Broadcasting the calls of a predator of the target species to deter the target species
Gas cannons and pyrotechnics pistols	Machines that produce a loud noise intended to scare animals away.
Ultrasonic/ infrasonic devices	Devices intended to deter animals by emitting noises audible to fauna, but above or below the human hearing range (i.e. below 20 kHz or above 20 kHz).
Acoustic road markings	Markings applied to roads that produce an artificial sound that deters animals when driven across.

Gas cannons and pyrotechnic pistols

Gas cannons and pyrotechnic pistols produce an artificial sound that has no social or other context. Gas cannons produce a loud bang by igniting gas (acetylene or propane). Pyrotechnic pistols produce sound from shells fired from guns emitting sound and light (e.g. flare). Some static gas cannons and pyrotechnic devices can be pre-programmed to fire at random intervals and fire in different directions; others can be remote controlled by an operator when required. Automatic devices can flush birds at inappropriate times (e.g. aircraft taking off/landing is an inappropriate time to haze birds at an airport). Continual maintenance of these devices can be onerous (Transport Canada 1998).

Ultrasonic/infrasonic devices

Ultrasonic/infrasonic devices emit an artificial sound that has no social or other context and is inaudible to humans, but audible to the target animals. Ultrasonic/infrasonic devices can be portable or static. Ultrasonic devices have been fitted to vehicles and trialled to prevent vehicle–fauna collisions (Bender 2001; Bender 2003; Muirhead *et al.* 2006). Ultrasonic devices have also been trialled on bats because they are able to hear ultrasound (Horn *et al.* 2008). Ultrasonic devices exist that are aimed at repelling insects, thereby reducing the food source of animals, such as bats, hence keeping them away or reducing their numbers at locations where they are at risk.

Acoustic road markings

Acoustic road markings are made from sound-producing material installed in bands along a road. The sound emitted when the markings are contacted by the tyres of passing vehicles is intended to deter fauna from roads (Ujvari *et al.* 2004). Acoustic road markings are not generally used in Australia to deter fauna.

Effectiveness

Because there are many different variations of auditory deterrents available, and they are all used in different situations (i.e. different environments to deter different species), the success or failure of specific devices cannot be generalised. There is, however, a growing consensus that auditory deterrents alone are not always effective over the long term at changing the behaviour of the target species. It cannot conclusively be said that auditory deterrents are ineffective in the absence of adequate experimental control and replication or without a detailed description of the deterrent and the situation in which it was applied. It is clear, however, that auditory deterrents alone should rarely be relied upon to deter fauna from a hazard at a development site. For this reason, auditory deterrents should be used on a trial basis where there is a lack of evidence that the deterrent has been effective elsewhere in a similar circumstance to where it is proposed to be used.

Recent studies on the use of distress, alarm and predator call deterrents in Australia are limited. Bender (2005) used a playback trial of a biologically significant signal – the foot thump of an eastern grey kangaroo (*Macropus giganteus*) – to investigate whether this sound could effectively deter other kangaroos within an agricultural setting. Kangaroos naturally produce the foot thump sound in response to a threat from a perceived predator and this signal is typically followed by flight. Bender's (2005) broadcasted foot thump sounds resulted in increased kangaroo vigilance levels and most individuals did indeed take flight. More recently, Biedenweg *et al.* (2011) suggested that incorporating artificial sound (a whip crack) with the biologically significant sound (the foot thump) may mask predictability and delay habituation.

Overseas studies have shown mixed results. Gilsdorf *et al.* (2004), for example, found that broadcasting recorded distress and alarm calls of deer triggered by an infrared detection system was ineffective at repelling deer in Nebraska. In contrast, Spanier (1980) tested distress calls to deter night herons (*Nycticorax nycticorax*) from trout ponds in Israel and concluded the calls were effective.

Biologically significant sounds such as distress calls are thought to be more resistant to habituation than artificial sounds such as blasts from gas cannons (Bomford and O'Brien 1990). However, their effectiveness needs to be assessed in each particular circumstance. Loud sounds produced by auditory deterrents, such as gas cannons and pyrotechnic pistols, may initially scare birds away from an area, but these devices have usually been found to be mostly ineffective in the long term because birds quickly habituate to them (see Transport Canada 1998; Donato *et al.* 2007). Habituation to auditory deterrents has been anecdotally reported to occur in a matter of days. These auditory deterrents that produce artificial sound with no social or other context are better used as part of an integrated hazing system (with consideration given to overcoming habituation, as described in Box 7.3). For example, an integrated hazing system (with radar-activated, on-demand deterrents (with gas cannons and predator calls) have successfully deterred waterfowl from using oil sands tailings ponds in Canada (Ronconi and St. Clair 2006).

There is mounting evidence that ultrasonic/infrasonic devices that also produce artificial sound that has no social or other context are poor at deterring Australian birds and mammals (Bomford 1990; Bender 2001, 2003; Magnus *et al.* 2004; Muirhead *et al.* 2006; Edgar *et al.* 2007). For example, Bender (2001) tested an ultrasonic device fitted to a vehicle and found that it did not alter the behaviour of eastern grey kangaroos or red kangaroos (*Macropus rufus*) and made no difference to the number of kangaroos struck by vehicles. Bender (2003) also tested a stationary ultrasonic deterrent and found that it did not reduce the density of free-ranging eastern grey kangaroos or change the behaviour of captive eastern grey kangaroos or red kangaroos. Similarly, Muirhead *et al.* (2006) found a stationary ultrasonic device did not deter the tammar wallabies (*Macropus eugenii*) living on Garden Island, Western Australia and Edgar *et al.* (2007) found a stationary ultrasonic device did not deter dingos (*Canis lupus dingo*) in captive trials. Overseas, there have been studies on the use of ultrasonic devices to repel bats, although results are inconclusive (Horn *et al.* 2008).

Overseas, acoustic road markings have reported to be poor at deterring fauna. For example Ujvari *et al.* (2004) found acoustic road markings were not likely to reduce the number of deer–vehicle collisions over the long term. Their potential use in Australia is also likely to be limited.

Design

The timing of broadcasting an auditory stimuli, as well as patterns and frequency of replication, is critical in creating the conditions that will likely be successful in repelling a target species. For example, it has been suggested that distress and alarm calls should be played before or as birds enter an area, rather than after they have arrived (Transport Canada 1998). When they are used, auditory deterrents need to be played sparingly to avoid habituation.

The acoustic envelope of an auditory deterrent needs to cover the area fauna are to be deterred from (Horn *et al.* 2008). Successful programs involving auditory deterrents will require diligence on the part of the managers to maintain the novelty and efficacy of the deterrent. It is likely that the most effective auditory deterrents are still waiting to be developed and will require a strong understanding of the behavioural tendencies of the target species. There is an onus on manufacturers and sale agencies to demonstrate, through third-party independent assessments, that devices will work in a given situation, particularly where high costs are involved.

Impacts on other aspects of the environment

Auditory deterrents are intended to deter native fauna from a hazardous component of a development site. Caution should be given when using auditory deterrents that radiate from a development site because their sounds may impact on nearby habitat areas. If a development is located in or near a noise-sensitive area (e.g. residential housing), auditory deterrents may not be suitable.

8

Measures to promote safe movement of fauna around and through a development site

Wildlife move across landscapes at different rates and for a variety of reasons, including dispersal, resource acquisition and to fulfil life cycle requirements. The movement of some species can be impeded or limited by tracts of cleared land between natural habitat and, it is for this reason, the maintenance of natural habitat linkages is often important for maintaining the long-term viability of plant and animal populations (e.g. van der Ree and Bennett 2003; Kirchner *et al.* 2003; Burns *et al.* 2004; Gilbert-Norton *et al.* 2010). It is becoming clear that effective protection, enhancement and creation of natural habitat linkages across the landscape is essential if we are to combat the negative effects of past land clearance and future land use.

Developments can present a barrier or otherwise impede fauna movement and flora dispersal. Habitat clearance and physical obstructions are the most common cause of barriers to movement. The degree to which a particular development fragments or isolates habitat will vary for each species. This is why an understanding of a species' mobility, dispersal mechanisms and/or movement in the landscape is required if we are to effectively apply measures to maintain movement. For example, arboreal mammals may have a dependency on particular movement corridors in highly fragmented landscapes (van der Ree *et al.* 2003).

As described in Chapter 4, avoiding impacts on natural habitat linkages is a primary consideration when planning a new development (e.g. re-positioning the development to avoid fragmenting habitat). In many cases this is not possible (e.g. highways linking towns across natural landscapes), so measures to promote the safe movement of fauna *around* and *through* development sites are used to increase the permeability of a development to wildlife movement. It may sometimes be more appropriate to facilitate movement of wildlife *around* a development due to hazards at the site.

Where complete exclusion of fauna is not required or desirable at a development site, a range of other strategies can be considered. The use of structures to promote the movement of fauna around and through development sites is a fast-growing area of environmental management. These structures include land bridges, fauna underpasses, arboreal rope bridges (Figure 8.1) and glide poles. Although these structures have been primarily developed to mitigate the adverse impacts of roads on wildlife, they are likely to have a wider application for other types of developments that impede wildlife movement/dispersal in a similar way. Many studies have been conducted on the use of movement structures overseas, but an understanding of how these measures can benefit Australian fauna is still limited. Forman *et al.* (2003) suggested six criteria for measuring the effectiveness of a crossing structure:

Figure 8.1. A ropeway integrated with glide poles over a land bridge at Hamilton Road, Brisbane (Photo: R Whitney)

- reduces rates of mortality
- maintains habitat connectivity
- maintains genetic interchange
- ensures biological requirements are met
- allows for dispersal and recolonisation
- maintains meta-population processes and ecosystem processes.

Fencing is a common component of various types of developments (e.g. residential and agricultural developments) and can impede fauna movement. This chapter describes how fences can be modified to be less intrusive for fauna moving through the area (i.e. inclusion fences and alternatives to barbed wire fences).

Natural habitat linkages

Overview	
What:	Arrangements of habitats in the landscape (wildlife corridors and stepping stones) that enables the movement of fauna and dispersal of flora.
Where:	Useful where developments present an opportunity to protect, enhance or create habitat linkages.
Pros:	Enables wildlife to disperse, acquire resources and fulfil life-cycle requirements. Maintains populations and genetic exchange.
Cons:	Species have particular requirements and may not use habitat links if they are not suitable (e.g. wide enough). In particular situations, wildlife corridors might potentially negatively impact animal populations by facilitating the movement of exotic animals, weeds, diseases and fire, increasing exposure to predators or human intervention and/or acting as dispersal sinks. Some wildlife may dominate a corridor and prevent its use by other species.

Cost:	Low–high, depending on whether existing habitat linkages are to be maintained or new habitat linkages are to be created.
Companion measures:	Canopy bridges, glide poles, land bridges and underpasses (all this chapter); revegetation and restoration of ecosystems, artificial frog ponds, wildlife-friendly dams, artificial tree hollows, artificial retreats for reptiles and salvaged habitat features (all Chapter 10).
Alternative measures:	Canopy bridges; glide poles (both this chapter).

Description and application

Natural habitat linkages are areas of habitat in the landscape that can be created or maintained to enable wildlife to move and disperse. They include wildlife corridors and stepping stones (described below). Many natural habitat linkages in the Australian landscape have been artificially created in a sense because the surrounding land has been cleared (e.g. natural habitat linkages are sometimes present as linear strips along roadsides in the Australian pastoral areas).

Impacts on remaining natural habitat linkages should be avoided to reduce impacts on species that use them for movement (Chapter 4) (Saunders and Hobbs 1991). If a natural habitat linkage is to be disrupted by development, it may be possible to enhance a degraded natural habitat linkage or create a new habitat linkage to maintain connectivity. Wildlife movement and/or dispersal capabilities for some species may be facilitated by:

- enhancing an existing link (e.g. widening a windbreak)
- establishing a new continuously linking wildlife corridor
- revegetating a block to provide a stepping stone
- establishing scattered trees (e.g. in an agricultural landscape).

The discussion on revegetation in Chapter 10 is applicable to the creation of a natural habitat linkage.

Wildlife corridors

A wildlife corridor (otherwise known as a 'habitat corridor' or 'greenway') generally describes a strip of vegetation that differs from the surrounding vegetation and connects otherwise separate areas of habitat (Saunders and Hobbs 1991). An example of a wildlife corridor could be a length of riparian vegetation flanking a creek between two larger vegetated areas in an agricultural landscape otherwise cleared of trees (Figure 8.2). Riparian zones with remnant fringing forest or woodland are often optimal corridors through a range of landscapes (Fisher and Goldney 1997; Bennett *et al.* 2000).

Wildlife corridors are not restricted to vegetation alone. For example, drainage ditches were shown to facilitate the movement of amphibians in Canada (Mazerolle 2004). The main function of a wildlife corridor is aptly described by Hobbs (1992) as a feature that allows wildlife movement from somewhere to somewhere else. However, this is not the only ecological function that corridors play. Other possible ecological values of wildlife corridors include intrinsic habitat (e.g. van der Ree and Bennett 2003; Pulsford *et al.* 2003), salinity management (e.g. Abel *et al.* 1997) and improved water quality (e.g. McKergow *et al.* 2006). Some potential disadvantages of corridors are discussed later in this chapter.

Natural habitat linkages are used in different ways by different species (Downes *et al.* 1997). Natural habitat linkages can act as habitat, filters, barriers, sources and sinks as illustrated in Figure 8.2 (Bennett 1991; Hess and Fischer 2001). Some species that avoid open areas may

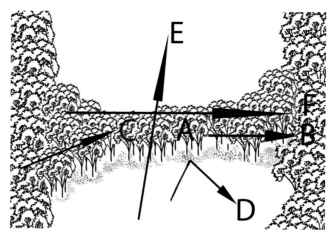

Figure 8.2. Ecological functions of wildlife corridors: (A) Habitat – organisms reside within the corridor; (B) Source – organisms within the corridor reproduce and spread outside the corridor; (C) Sink – organisms enter the corridor and perish; (D) Barrier – organisms from the surrounding landscape cannot pass through the corridor; (E) Filter – only select organisms can pass through the corridor; (F) Conduit – organisms use the corridor for movement

particularly benefit from the provision of a corridor with a reduced area-to-edge ratio (e.g. small forest bird species).

Wildlife corridors operate at various spatial scales: local corridors join particular habitat areas tens of metres apart, while regional corridors span hundreds of kilometres (Pulsford *et al.* 2003). Wildlife corridors are not always comprised of woodland or forest. Open grassland, for example, may assist the movement of the green and golden bell frog (*Litoria aurea*) between wetlands (Burns *et al.* 2004; Muir 2008), thus effectively acting as a corridor assisting movement.

Stepping stones

One or more disjunct blocks of suitable habitat between particular habitat areas can also provide a conduit for movement for some species and, in this regard, the habitat patches are generally referred to as 'stepping stones' (Date *et al.* 1991; Lindenmayer and Fischer 2006) (Figure 8.3). Vegetated stepping stones are used by various birds and bats. Small patches of vegetation in the landscape are also important sources of seed for regeneration of adjacent

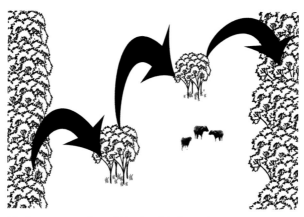

Figure 8.3. A graphical representation of blocks of habitat that act as stepping stones between larger habitat areas

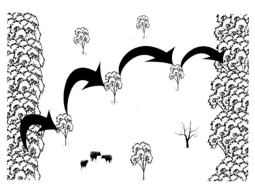

Figure 8.4. A graphical representation of scattered trees that act as stepping stones between larger habitat areas

vegetation. For example, unmined remnant patches of vegetation surrounding mined areas can assist with recolonisation of flora to rehabilitated land. Other habitats, such as dams, can also act as stepping stones for fauna (Brainwood and Burgin 2009).

Patches of vegetation do not need to be large to provide some benefit to wildlife. Some species (e.g. grey-crowned babbler) use scattered trees such as those in an agricultural landscape (Figure 8.4). Scattered trees in the Australian landscape are important to native fauna, not only for movement but also for sources of foraging and nesting resources (Fischer and Lindenmayer 2002; Manning *et al.* 2006; Gibbons *et al.* 2008; Fischer *et al.* 2009). In some situations, they mimic open woodlands, providing habitat for a range of species.

Effectiveness

Natural habitat linkages have been shown to be effective for the dispersal of some flora (e.g. Damschen *et al.* 2006) and many fauna species, such as the short-billed black-cockatoo (*Calyptorhynchus latirostris*) (Saunders and Ingram 1987), and gliders (*Petaurus* spp.) (van der Ree *et al.* 2003). However, whether a particular natural habitat linkage is likely to be effective should be evaluated on a case-by-case basis because different species–environment interactions will influence the effectiveness of particular linkages (Simberloff and Cox 1987; Harris and Scheck 1991; Beier and Loe 1992; Soule and Gilpin 1991; Beier and Noss 1998).

Design

There is limited information for the minimum design requirement for maintaining or creating effective natural habitat linkages for particular species in Australia. There is no minimum effective corridor width or stepping stone patch size that can be generically applied to all different scenarios because they will vary with species, time, habitat and landscape. Table 8.1 includes some factors that influence the effectiveness of wildlife corridors and stepping stones that facilitate movement for a range of species.

While it is generally true that wider corridors are used by a wider variety of species, narrow wildlife corridors (5–10 m in width) can sometimes provide positive outcomes for wildlife movement (Lynch and Saunders 1991; Hellmund and Smith 2006; Kinross and Nicol 2008). Narrow wildlife corridors occur naturally in arid areas of Australia as riparian trees surrounded by arid vegetation. In disturbed environments, however, narrow corridors can also advantage habitat generalists (e.g. aggressive noisy miners, *Manorina melanocephala*, that use a variety of habitat types) to the detriment of smaller native birds with more specific habitat requirements, such as honeyeaters, robins and thornbills (Buchanan 2009).

The effectiveness of stepping stones in promoting movement of species will depend on the spacing of the habitat patches or trees in relation to the movement of the species. The habitat

Table 8.1.   Examples of factors that influence the effectiveness of wildlife corridors and stepping stones

Component	Factor	Example
Areas to be connected	Composition of the habitat	Habitat type, quality and diversity should suit the target species so the species is able to use the areas to be connected.
	Control of threatening processes	Threatening processes, such as weed and pest species, may need to be controlled so that they do not spread once the areas are connected.
	Presence of competing species	Additional habitat resources (e.g. artificial tree hollows) may need to be provided in order to accommodate species moving into the connected habitat.
Natural habitat linkage	Composition of the wildlife corridor or stepping stone	Habitat type, quality and diversity should suit target species so the species are able to use the wildlife corridor or stepping stones.
	Width of the wildlife corridor or stepping stone	A narrow wildlife corridor (e.g. tens of metres) may provide a conduit for movement for some species; however, a wider wildlife corridor (e.g. hundreds of metres) would provide greater habitat complexity and be used by a wider variety of species.
	Length of the corridor	The longer the distance between habitats the wider the linkage likely to be required for forest interior species that are less likely to cross long, narrow linkages.
	Presence of competing species	Additional habitat resources (e.g. artificial tree hollows) may need to be provided in the corridor to overcome issues with competing species within the corridor.
	Gaps in the linkage	Gaps in the wildlife corridor may be barriers to the movement of some species but inconsequential for other species.
Surrounding landscape	Influences in the surrounding landscape	The natural habitat linkage and areas it links should have a reasonable prospect of remaining suitable habitat in the future.
	Presence of habitat resources in the surrounding landscape	A wildlife corridor may be used by more fauna if there are other habitat resources in the surrounding landscape, such as scattered trees, as opposed to cleared land.
	Presence of competing species	Exotic animal control may be required on surrounding land to minimise predation on species using the corridor.

resources provided by individual scattered trees may also influence its usage by fauna (e.g. flowering eucalypts).

Supplementing habitat resources within a natural habitat linkage with artificial tree hollows, artificial retreats for reptiles and salvaged habitat features (Chapter 10) may also increase its usage by wildlife.

Placement

Many of the design factors described above will also affect where new habitat linkages are created and the timeframe in establishing an effective linkage. Many of the factors affecting revegetation and restoration, as described in Chapter 10, are also applicable. For example, regenerating woodland to provide a habitat linkage is likely to be easier if it is located within land with medium to high resilience, considering factors such as historical land use and soil properties (Keydoszius *et al.* 2007).

Its location in the catchment will influence its effectiveness and how easily the natural habitat linkage can be established. High priority should be given to maintaining or creating wildlife corridors along streams and watercourses because these types of natural linkages can form an interconnected system through catchments, provide rich habitats for wildlife and can have other ecological benefits (e.g. soil stabilisation) (Fisher and Goldney 1997; Bennett *et al.* 2000). Natural linkages should generally be created to join habitats that were previously linked. Connecting historically separated habitats may disrupt naturally isolated populations (Saunders and Hobbs 1991).

Maintenance

The need for management (e.g. weed control) of linear strips of habitat is generally higher than the management of blocks of vegetation, primarily due to increased edge effects (Loney and Hobbs 1991). In this regard, wider corridors are easier to manage (Saunders and Hobbs 1991).

For the future persistence of scattered trees, it is important to have younger regeneration trees in the mix to replace the older trees once they die. Fencing around groups of scattered trees in livestock paddocks can assist the regeneration of vegetation between them, thus forming larger stepping stones.

Impacts on other aspects of the environment

Where there is uncertainty over the benefits of a natural habitat linkage, the proposed corridor(s) should be assessed by an experienced ecologist because simplistic application of general principles can do more harm than good (Soule and Gilpin 1991). Various authors have highlighted that the risk and cost of providing new habitat linkages may outweigh the ecological benefits (Simberloff and Cox 1987; Hobbs 1992). It has been noted that, in some situations, wildlife corridors can have a negative impact on animal populations by facilitating the movement of pests, weeds, diseases and fire, acting as sources of pests, weeds and diseases, increasing exposure to predators, increasing susceptibility of impacts from humans and/or acting as ecological sinks (Simberloff and Cox 1987; Saunders and Hobbs 1991; Hilty *et al.* 2006). However, possible adverse affects should not always deter natural habitat linkages from being established. Instead, forward planning should identify possible adverse affects so measures can be put in place to avoid or minimise them.

Land bridges

Overview	
What:	A bridge structure, topped with soil and vegetation, to facilitate the movement and dispersal of wildlife (mainly animals) across an otherwise dangerous or impenetrable barrier.
Where:	Where a development (e.g. roads, railway lines, canals or pipelines) is to be spanned to provide continuity of habitat or to facilitate movement of animals.
Pros:	Links fauna habitat; facilitates movement of animals; traffic related mortality is reduced when used in conjunction with exclusion fencing.
Cons:	Some species may not use a land bridge.
Cost:	High.
Companion measures:	Fauna exclusion fencing (Chapter 7); natural habitat linkages, canopy bridges and glide poles (this chapter).
Alternative measures:	Fauna underpasses, canopy bridges and glide poles (all this chapter).

Description and application

Land bridges (sometimes known as 'fauna overpasses', 'green bridges', 'ecoducts', 'wildlife bridges' or 'landscape connectors') are bridge structures topped with soil and vegetation that can be used to facilitate the movement of fauna and the dispersal of flora across an otherwise dangerous or impenetrable barrier. Although 'fauna overpass' is the most common name given to this crossing structure overseas, we have chosen to refer to them as land bridges here to prevent confusion with two other types of overpasses installed in Australia – glide poles and canopy bridges (both discussed later in this chapter). Land bridges are worth considering to promote the safe movement of some fauna across linear infrastructure developments because they provide a seemingly natural continuum of habitat across the bridge. Land bridges are intended to maintain or re-establish connectivity of habitat at a broader landscape level. In particular, land bridges have been used to reduce traffic-related mortality.

Although land bridges are mainly used over roads, there is the potential to use the concept in association with other developments such as railway lines, canals and pipelines. Bridges of land can be incorporated into post-mine landscapes to maintain linkages across otherwise expansive open cut pits. Although this is not applicable to all mining developments, it will become increasingly important to maintain linkages between habitats in mining regions where altered landscapes may present barriers to species movement.

Land bridges are sometimes specifically constructed to benefit a particular species such as a threatened species or one that is prone to vehicle strike. Land bridges would generally be expected to benefit numerous species if designed well.

The design and placement of a land bridge should encourage natural faunal movements (e.g. movements within a home range, migration and/or dispersal) and promote movement between existing habitats. Most existing land bridges are around 30–50 m wide (Forman *et al.* 2003) and are generally covered with soil and mulch, planted with local flora species and furnished with rocks and logs to provide shelter and protection from predators for animals.

Fauna exclusion fencing (Chapter 7) is required alongside roads and over land bridges for the purposes of preventing animals from falling from the bridge or otherwise entering the roadway, as well as channelling them towards the land bridge. Without this 'channelling', some species may not find the entrance to the land bridge.

Most existing land bridges in the world have been constructed in Europe. France was the first European country to use land bridges, with 125 in place by 1991 (Bank *et al.* 2002). Land bridge usage in Australia is in its infancy, but it is becoming increasingly common for them to be incorporated into new road designs in the eastern states (Queensland and New South

Figure 8.5.   Land bridges constructed over the Yelgun to Chinderah Freeway section of the Pacific Highway in northern New South Wales at (a) Tagget's Hill near Cudgera Creek and (b) Marshall's Ridges near Yelgun (Photos: D Gleeson)

Figure 8.6.   A land bridge constructed over Compton Road, southern Brisbane (a) a side view and (b) a view from the top (exclusion fencing demarcates the edges). Glide poles are erected along the length of the land bridge (Photos: D Gleeson)

Wales). Examples of land bridges in Australia can be seen at the Yelgun to Chinderah Freeway section of the Pacific Highway in northern New South Wales (Figure 8.5), at Compton Road, southern Brisbane (Figure 8.6) and at Hamilton Road, Brisbane (Figure 8.7).

van der Ree *et al.* (2008a) suggested that a separate classification of land bridge be given to those that incorporate linear infrastructure that typically allow human access. This type of dual-purpose land bridge often incorporates minor roads, but can also be used by fauna. An Australian example of this type of structure can be seen spanning Caloundra Road on the Sunshine Coast in Queensland (Figure 8.8).

Land bridges are generally one of the most expensive artificial wildlife crossing structures due to the high engineering and construction costs associated with their large size. Land

Figure 8.7.   A land bridge constructed over Hamilton Road, Brisbane (note integration with glide poles and canopy bridge) (Photo: R Whitney)

Figure 8.8.   A dual purpose land bridge constructed over Caloundra Road, Sunshine Coast (note integration with canopy bridge) (Photo: R Whitney)

bridges are used less often than underpasses worldwide (van der Ree *et al.* 2008a), probably due to cost. Huijser *et al.* (2009) estimated that the cost of a land bridge is around ten times higher than a fauna underpass (discussed later in this chapter). Inclusion of land bridges at the planning stage of road design is generally more economical than retrofitting them into existing roads. Careful design is required because the costs of installing and maintaining land bridges is substantial and unwarranted if they do not fulfil their purpose.

One innovative way of overcoming cost constraints in the future may be to use modular, plastic land bridges that can be quickly assembled where needed (Huijser *et al.* 2010).

Effectiveness

The installation of land bridges to mitigate the impacts of roads in Australia has outpaced the research on the efficacy of the structures, particularly with regards to benefits for wildlife populations. The research conducted by Bond and Jones (2008) on the land bridge in Brisbane (Figure 8.6) and by Hayes and Goldingay (2009) on the two land bridges constructed in northern New South Wales (Figure 8.5) are the only examples of peer-reviewed published data describing the faunal usage of land bridges in Australia to date.

Like many studies worldwide, Bond and Jones (2008) and Hayes and Goldingay (2009) clearly demonstrated that various types of animals will use land bridges. van der Ree *et al.* (2008a) undertook a comprehensive literature review of Australian land bridges and reported use by the following native species: eastern blue-tongue (*Tiliqua scincoides*), lace monitor (*Varanus varius*), short-beaked echidna (*Tachyglossus aculeatus*), eastern grey kangaroo (*Macropus giganteus*), spotted-tailed quoll (*Dasyurus maculatus*), red-necked wallaby (*Macropus rufogriseus*) and swamp wallaby (*Wallabia bicolor*). Other native fauna were recorded using the land bridges but were not identified to species level, including ducks, herons, bandicoots, possum, rat, skinks and snakes and unnamed smaller mammals (van der Ree *et al.* 2008a). It is likely that many more Australian species will be found to use land bridges in the future.

Not all species will use land bridges – animals such as bandicoots and rodents prefer underpasses (Hayes and Goldingay 2009), which are discussed later in this chapter. Such preferences seem to correlate with their evolved behaviours and life history traits (i.e. some species prefer open areas and others need cover). The effectiveness of a dual-purpose land bridge is unknown, but because they may be avoided by shy fauna, their effectiveness for a wide range of species is questionable.

At the time of writing, no published studies had examined the effects of roads on gene flow of fauna populations in Australia (Taylor and Goldingay 2010) or the effects of land bridges at facilitating gene flow. Various overseas studies have revealed that land bridges would aid in maintenance of gene flow. For example, Olsson *et al.* (2008) suggested that five to seven moose (*Alces alces*) per year using a land bridge would maintain gene flow between two otherwise separate sub-populations. As yet, there is no evidence that land bridges are effective at preventing genetic isolation (Corlatti *et al.* 2009).

Design

Examples of factors to be considered when designing a land bridge are described in Table 8.2. It should be kept in mind that different solutions will be needed for different species.

Placement

Land bridges are generally used in conjunction with fauna exclusion fencing (Chapter 7). Due to the potential negative impacts of fencing, land bridges should only be used along roads with

Table 8.2.   Some design considerations that influence the effectiveness of a land bridge

Consideration	Description
Fauna exclusion fencing	Fencing helps prevent animals from entering the roadway, as well as to funnel individuals onto the land bridge.
Width	Adopting a greater width may increase the effectiveness of a land bridge, but several smaller land bridges may be more effective than one large land bridge (Bank *et al.* 2002).
Depth of soil	Soil depth affects hydrology and the vegetation that will be able to be established. It has been suggested that a minimum of 300 mm of soil is needed to establish grasses, 600 mm to establish shrubs and 1.5–2 m to establish trees (Queensland Department of Transport and Main Roads, 2010). However, even soil at this depth may not be sufficient to support the growth of larger trees.
Ground cover	The cover of ground vegetation across a land bridge should generally be comparable to that in the habitats either side of the land bridge. Ground cover is important habitat for both invertebrate and vertebrate species.
Vegetation	The vegetation on a land bridge should ideally be similar in floristics and structure to adjacent vegetation to provide a continuum of habitat (e.g. Jones *et al.* 2011). This is likely to encourage species naturally occurring on each side of the land bridge to use it.
Furniture	'Furniture' may increase the effectiveness of a land bridge, depending on the habitats to be connected and the species likely to use it. Furniture may include rocks and logs.
Disturbance	Vehicle traffic volume may affect use by some species. Berms or board fences can be placed along the sides of the land bridge to minimise traffic noise and lights disturbing fauna.
Approaches	Suitable habitat leading to, and occurring near, the land bridge encourages use by fauna.

Box 8.1.   Vegetated overpasses assist small birds to cross roads

Associate Professor Darryl Jones, Environmental Futures Centre, Griffith University

Most purpose-built fauna passages have been designed and constructed with terrestrial mammals in mind. Unsurprisingly, birds have been largely ignored, presumably because they can fly. Although a busy road will present a significant risk or complete barrier to the movements of a snake, bandicoot or wallaby, we tend to assume that most birds will simply take to the air in order to cross. This is certainly the case with the majority of larger species of birds, as well as those that live in open habitats. Recently, however, a range of studies from around the world have confirmed that many birds that live in forested landscapes avoid major habitat gaps, and are often reluctant to fly across even modest spaces; studies from South America found a gap of only 45 m to be a critical threshold. For species such as these, where even a natural watercourse represents a complete barrier, the existence of a road – or any form of linear clearings – will often be a major impediment to normal movements. Furthermore, many species also avoid habitat edges and will tend to retreat further into the depth of the undisturbed interior of the forest.

These recent findings have significant implications for the impact of roads on populations of many forest bird species. With the ever-expanding road network increasingly fragmenting forested habitats into smaller and smaller segments, the long-term survival of smaller forest birds becomes of considerable concern. One measure that could potentially encourage the safe passage of birds over roads may be fauna overpasses. Unfortunately, very little information is currently available on bird use of these structures. Moreover, because they were designed primarily for larger mammals, they tend to be sparsely vegetated and typically placed in open environments – conditions unlikely to attract the forest birds of concern.

Australia currently has five fauna overpasses – two in southern Queensland and three in northern New South Wales. Although constructed primarily for macropods – kangaroos and wallabies – all of these structures were also designed to be used by a wider diversity of species (e.g. two include a series of glide poles) and all were intended to provide habitat continuity between the forested landscapes on either side of the road. To this end, all of the Australian fauna overpasses are well vegetated.

By far the best studied of these structures is that spanning almost 100 m of the four-lanes of Compton Road in southern Brisbane. Completed in 2005, and planted with many thousands of plants of local provenance, by 2010 the vegetation closely resembled the subtropical eucalyptus forest present on either side of the road. Intensive studies of the birds using this structure commenced in 2008, when the planted understorey vegetation had developed into a dense, continuous layer. To the researcher's considerable surprise, not only were significant numbers of birds detected on the overpass – almost all within the dense understorey vegetation – the diversity of species was remarkably rich. To date (early 2011) 30 species have been recorded using the structure, most of which were smaller species (the modal size being 12 g, which is typical of a honeyeater). Of particular significance, however, the species detected included treecreepers, flycatchers and robins, which are typically regarded as being especially edge-adverse and sensitive.

These early findings are potentially of great importance because they suggest that the most vulnerable smaller species of birds, even the more sensitive, will use overpasses to cross wide, busy roads. However, we believe that the presence of dense vegetation cover, preferably of local species, is critical to the success of these measures.

high traffic volumes where fauna–vehicle collisions are most likely to occur or if there is a species of concern (e.g. a threatened species) that is likely to be adversely impacted by the presence of a roadway.

To improve the effectiveness of a land bridge, it should be placed in an area of suitable habitat and within the natural paths of target species. The surrounding land use and tenure should be considered when planning a land bridge. The habitat to be connected by the land bridge should have a reasonable prospect of remaining suitable habitat in the future. Any obstruction to animals being able to access the land bridge (e.g. fences and irrigation channels) would reduce the likelihood that animals will use the bridge. A lack of vegetation at the entrance of a vegetated land bridge (e.g. as a result of clearance) in some instances will make the land bridge unsuitable for some species, such as small forest birds (Box 8.1; see also Jones and Bond 2010).

Maintenance

Frequent inspections and maintenance of a land bridge is important. Maintenance involves mulching, removal of excess dead wood, remediation of soil erosion and maintaining vegetation cover.

Impacts on other aspects of the environment

Although land bridges can aid in the movement of animals across linear infrastructure, the effects on populations under a range of scenarios is largely unknown. For example, could the land bridge siphon animals away from the core population, if they move across the bridge and cannot return? Alternatively, if land bridge links a large patch of habitat to a small patch of habitat, will the animals choose to use only the larger patch?

Exotic animals were also recorded using the Australian land bridges including the cane toad, cat, cow, dog, fox, horse and brown hare (van der Ree *et al*. 2008a). Exotic animal control may be required to minimise the spread of exotic animals and increase the use of the land bridge by native species.

Fauna underpasses

Overview	
What:	A structure that can facilitate the safe movement of some fauna underneath a development.
Where:	Beneficial when a development (e.g. roads and railway lines) needs to be traversed to provide continuity of habitat or facilitate movement.
Pros:	Links fauna habitat; facilitates movement; traffic related mortality is reduced when used in conjunction with exclusion fencing.
Cons:	Some species may not use a fauna underpass. Underpasses rarely replace all of the attributes of the existing movement corridor.
Cost:	Medium–high. Small culverts are reasonably economical to install, whereas large underpasses tend to be expensive due to construction costs associated with their large size.
Companion measures:	Fauna exclusion fencing (Chapter 7); natural habitat linkages (this chapter).
Alternative measures:	Avoid impacts to fauna movement (Chapter 4); land bridges (this chapter).

Description and application

Fauna underpasses (sometimes known as 'wildlife tunnels') are structures intended to facilitate the safe movement of some fauna under developments, such as roads and railway lines, by allowing animals to travel under infrastructure. The design and placement of a fauna underpass should aid natural faunal movements (e.g. movements within a home range, migration and/or dispersal) and promote habitat connectivity.

It is not uncommon for animals to move through pre-existing structures built for other purposes. For example, drainage culverts and small below-grade access roads are commonly used as passageways under infrastructure by fauna (e.g. Grilo *et al.* 2008). Purpose-built wildlife crossing structures have been used widely in North America and Europe for decades, but are now gaining popularity for use in Australia to connect forest remnants or other habitats divided by roads. Inclusion of underpasses at the planning stage of road design is usually more economical than retrofitting them into existing roads.

Fauna underpasses can be dry or can contain water (e.g. placed in locations where water is required to drain under roads), depending on the species that are intended to use them. Fauna underpasses can be made more attractive to fauna by including additional habitat components termed 'furniture', such as logs, ledges (raised shelves) and rocks. For example, hollow logs and pipes placed within underpasses may encourage use by smaller mammal species.

Purpose-built fauna underpasses were first used in Australia in the 1980s for the dispersal of the mountain pygmy-possum (*Burramys parvus*) at Mount Higginbotham in Victoria (see Mansergh and Scotts 1989). Fauna underpasses range from small structures based on pipe culverts through to large passageways under bridges/viaducts (Table 8.3). Different fauna will use different sized culverts.

Small fauna underpasses

Small fauna underpasses can be installed in areas likely to benefit the movement of smaller animals such as frogs. They are often made from pipe culverts of concrete or metal with a narrow diameter. An example of a small fauna underpass installed for use by the wallum sedge frog (*Litoria olongburensis*) along the Tugun Bypass in South-East Queensland is shown in Figure 8.9.

Medium fauna underpasses

Medium fauna underpasses are often made from box culverts of concrete or metal with a large diameter. An example of an underpass made from a box culvert is shown in Figure 8.10. This fauna underpass was constructed under Compton Road, southern Brisbane (see Box 8.2) and is 2.4 m high, 2.5 m wide and 48 m long (Bond and Jones 2008).

Large fauna underpasses

Large fauna underpasses are passageways beneath bridges or viaducts. In this situation, the underpass is the valley beneath the bridge or viaduct. Because they are generous in size, these

Table 8.3. Types of fauna underpasses

Type	Description
Small fauna underpasses	Tunnels generally made from narrow pipe culverts. Also referred to as 'amphibian tunnels'.
Medium fauna underpasses	Tunnels generally made from box culverts or similar.
Large fauna underpasses	Large, open passageways formed beneath bridges/ viaducts.
Converted structures	Existing drainage culverts modified to encourage fauna to use as an underpass.

Figure 8.9. A small fauna underpass installed along the Tugun Bypass in South-East Queensland. Note the water in the culvert improves its attractiveness to frogs (Photos: J Gleeson)

underpasses can accommodate a wider range of species and enable the movement of large fauna with little restriction. Examples of large fauna underpasses include bridges constructed over riparian habitat flanking watercourses and viaducts that allow traffic to flow high above the valley floor. In both examples, animals are free to move under the infrastructure as if it were not there (although traffic noise may discourage some species). Large fauna underpasses are particularly suited to mountainous areas because the natural variation in topography can be exploited. Examples of remarkable viaducts that also function as large fauna underpasses can be seen overseas in Europe (e.g. Viaducto de Montabliz in Spain).

Figure 8.10. Approach to a medium fauna underpass at Compton Road, southern Brisbane. Note the log railing installed as a 'walkway' (Photo: D Gleeson)

Converted structures

In some instances, it is possible to inexpensively modify pre-existing structures built for other purposes (e.g. drainage culverts and small below-grade access roads) to make them suitable for use by fauna, such as planting near the entrances so that animals approaching the underpass feel safe from predators. Dry ledges, raised shelves or railings can be installed above areas subject to water inundation so that animals that do not swim (e.g. small and medium mammals) can use the structure as an underpass.

Effectiveness

Studies conducted in Australia and overseas have confirmed that underpasses are used by a wide range of fauna; however, species differ in their willingness to enter or pass through a fauna underpass (Norman *et al.* 1998; Glista *et al.* 2009) (Box 8.2). The use of underpasses by fauna has generally been investigated using observational techniques such as detection of animal tracks in a suitable medium such as sand and/or using cameras (Goosem 2005; Ford *et al.* 2009). Overseas, Woltz *et al.* (2008) used an experimental approach to determine that small underpasses made of round PVC pipe of at least 0.5 m diameter and lined with soil or gravel are used by the frog and turtle species investigated when installed in tandem with a 0.6–0.9 m high fauna exclusion fence.

Fauna exclusion fences (Chapter 7) are generally used in conjunction with fauna underpasses to prevent animals from entering the development (e.g. roadway) and to help funnel individuals into the fauna underpass. Bond and Jones (2008) found that fences used in association with underpasses prevented most road-kill, especially of larger animal species, except where breached by vandals. Fauna underpasses are less effective without a fence to funnel the fauna because they would only be used if they were located within a pre-existing fauna movement pathway.

It is important to know whether a fauna underpass provides benefits to populations (Clevenger and Sawaya 2010) particularly with regard to sustaining the long-term viability of a local population (van der Ree *et al.* 2007) (Box 8.3). Very few studies have compared population levels 'before' and 'after' the installation of wildlife crossing structures.

In the mid 1980s, Mansergh and Scotts (1989) demonstrated that the dispersal of the mountain pygmy-possum (*Burramys parvus*) at Mount Higginbotham in Victoria was maintained using medium purpose-built underpasses (rock-lined box culverts 0.9 m × 1.2 m). More recently, van der Ree *et al.* (2009) investigated the effectiveness of this underpass using population viability analysis (i.e. comparing rates of birth, death and migration before and after the construction of the road and underpass). Using a 20-year dataset, they found that the underpasses reduced, but did not entirely eliminate, the negative effects of a road for the mountain pygmy-possum. It is worth noting that this was perhaps the first study to investigate the effectiveness of an underpass using population viability analysis.

Most agencies in the US that use underpasses and fences to reduce collisions with fauna have reported that they were effective (Glista *et al.* 2009). Fauna underpasses provided in association with fauna exclusion fencing are likely to be generally more effective in preventing road-kill than measures that rely on people to voluntarily change their behaviour (e.g. signs recommending lower traffic speed) (Chapter 5).

Design

The type of fauna underpass (small, medium or large) will need to suit the species proposed to use it. Underpasses should be designed with features to encourage fauna use, such as natural-looking approaches and substrate. Examples of factors to be considered when designing a

Box 8.2.   Compton Road wildlife underpass

Compton Road, a major arterial located south of the Brisbane metropolitan area, has one of the largest concentrations of wildlife crossing structures in a single location anywhere in the world, including a land bridge, fauna underpass, glide poles and canopy bridges (Bond and Jones 2008). Their structures were considered necessary because the road was upgraded from two to four lanes along the section that separates the nationally significant Karawatha Forest from Kuraby Bushland to the north.

The wildlife crossing structures include two fauna underpasses and three wet culverts. Each fauna underpass is 2.4 m high, 2.5 m wide and 48 m long (Bond and Jones 2008). The approach to, and entrance of, one of the underpasses is shown in Figure 8.10. Each underpass comprises three levels. The lowest level is made of concrete and facilitates water flow. A middle level, also made of concrete, is equipped with rocks. The highest level consists of two types of shelves that run the full length of the underpass. One of these shelves is a narrow, flat, wooden shelf attached to the wall of the underpass and the other is a railing made from logs. The interior of one of the underpasses is shown in Figure 8.11.

Figure 8.11.   Internal design of one of the fauna underpasses at Compton Road, southern Brisbane (Photo: D Gleeson)

Bond and Jones (2008) found that a wide range of species, especially small to medium-sized mammals, visited the fauna underpasses. In fact, fauna began visiting soon after construction was completed despite the large distance between the underpass entrance and the existing forest. The most frequent visitors to the underpasses were small mammals, mostly rodents, as well as a large number of bandicoots. Cats and dogs, both known to prey on native fauna, also used the underpasses. The raised shelves were used by many small mammals (20% of rodents and 40% of dasyurids entered the underpass travelling above the floor).

Substantial numbers (approximately 16–30%) of individuals entering the underpass made the entire crossing from one end of the underpass to the other. Possums, bandicoots, cats, dogs and frogs were more likely to traverse the entire length of the underpass. Reptiles were the second most abundant taxon recorded during Bond and Jones' (2008) study, but only a few appeared to make the full crossing from one side to the other.

Box 8.3.   Integrating road underpasses for the threatened growling grass frog (*Litoria raniformis*) into a broad-scale residential development in south-east Melbourne

Dr Andrew Hamer, Australian Research Centre for Urban Ecology
Dr Aaron Organ, Ecology Partners Pty Ltd

The construction and use of new roads in the Pakenham area represents a significant threat to the growling grass frog (also known as the southern bell frog) population, because road construction removes habitat and has direct and indirect impacts to nearby habitats (Forman and Alexander 1998; Trombulak and Frissell 2000), and road traffic can kill frogs while they are crossing the road (Fahrig *et al.* 1995; Hels and Buchwald 2001) (Figure 8.12). The impacts of roads on individual frogs and their habitat can extend to negative effects at the population level, which may not become evident for decades, and may ultimately cause local extinction (Findlay and Bourdages 2000). In the Melbourne region, high road density is known to reduce the number of frog species occupying a pond (Parris 2006). One method of mitigating the impact of roads on frog populations is the inclusion of culverts or tunnels ('underpasses') and drift fences that facilitate the movement of individuals under the road. These structures have been shown to be used by frogs and toads in Europe and North America (Dodd *et al.* 2004; Lesbarrères *et al.* 2004). However, in Australia the only amphibians known to regularly use underpasses are cane toads, and culverts have not been shown to mitigate road-kills on native frogs (Taylor and Goldingay 2003).

Figure 8.12.   Growling grass frog (*Litoria raniformis*) (Photo: A Organ)

Ten underpasses have been installed into the Pakenham Bypass, which is a 20 km six-lane road constructed in 2008, specifically for the growling grass frog (Figure 8.13). A series of 32 ponds were also created to offset the loss of farm dams during construction, and were located both at the entrances of underpasses and along the bypass right-of-way close to underpasses. These measures were a requirement of the Victorian Minister for Planning granting project approval, and monitoring of the population and its habitat were also a requirement under the decision by the

Commonwealth Department of Sustainability, Environment, Water, Population and Communities under the Commonwealth *Environment Protection and Biodiversity Conservation Act 1999*. The placement of underpasses and creation of wetlands directly adjacent to these water bodies was guided by the results of previous surveys for the southern bell frog in the area (Organ 2004; Organ and Hamer 2006), and were strategically positioned in close proximity (i.e. within 500 m) of occupied sites, or sites that had a moderate to high likelihood of future occupancy. Underpasses were also installed at points along the bypass that intersected natural drainage lines, which also have the potential to act as underpasses, because it is hypothesised that drainage lines act as movement corridors for this species (Hamer and Organ 2008). Protocols for monitoring the population were formalised in a Conservation Management Plan produced for the proponent (Organ 2005). The underpasses were deemed necessary to facilitate the movement of growling grass frogs under the bypass, thereby maintaining connectivity between water bodies and reducing the potential of the road to act as a barrier to frog dispersal. Drift fences engineered to a 'frog-proof' design were installed either side of the bypass to prevent frogs moving onto the road, and to direct frogs towards underpasses. Habitat for the growling grass frog was created at the entrances to underpasses through the construction of permanent wetlands containing key habitat attributes required by the species, including a diversity of aquatic and semi-aquatic vegetation, and surrounded by terrestrial cover such as rocks and logs.

Figure 8.13. Underpass and created wetland, Pakenham Bypass, Victoria (Photo: A Organ)

The monitoring program was initiated in 2005 and included annual field surveys for the growling grass frog during the spring and summer breeding season. Water bodies were surveyed up to 1 km ('road-effect zone'; Eigenbrod *et al.* 2009) of the bypass and included newly created wetlands and existing farm dams. Adult frogs that were captured during the surveys at a pond were uniquely marked to determine if frogs had moved from one side of the road to the other. Monitoring of underpass wetlands commenced after the construction of the Bypass in 2008 and, approximately a year and a half after wetland creation, frogs colonised 39% of created water bodies, with successful breeding and recruitment documented at 5% of these sites. In the second year of monitoring, frogs colonised 23% of created water bodies, with successful breeding and recruitment documented at 5% of these sites. Although most inter-annual frog recaptures were from the wetlands where they were initially marked, one individual in 2010 was recorded in a wetland on the opposite side of the road 9 months after the initial capture. However, it is assumed that the frog moved under the road along a creek crossing, and had not trespassed the drift fence and moved across the road surface. This marks the first recorded north–south movement across the Bypass since the commencement of monitoring. It is likely that the frog moved underneath the Bypass via Toomuc Creek. Another individual in 2010 was recaptured in a pond on the same side of the road approximately 200 m away, and there have been similar short distance movements documented between wetlands on the same side of the road.

Although mitigation measures such as crossing structures have the potential to maintain road permeability for range of species including amphibians, at this stage the monitoring results are not conclusive to determine whether these structures are adequate to cater for population dynamics of the growling grass frog. Heard *et al.* (2010) assert that, because of the experimental nature of underpasses, they should not at present be regarded as primary measures to mitigate or offset the impacts of urban development on populations of the growling grass frog. Accordingly, potential impacts of the Bypass on the growling grass frog must continue to be monitored at the population level in Pakenham/Officer, and created and extant habitat in the vicinity of the Bypass needs to be managed for this species over the long term.

fauna underpass are described in Table 8.4. It is likely that there is a maximum or optimal length of a fauna underpass that a particular species is more likely to use, but this has not yet been determined for Australian fauna. In any case, the length of an underpass is often dictated by the design of the development (e.g. the width of a road).

Studies have shown that the furniture and substrate inside an underpass can increase usage by some fauna. For example, some 20% of rodents and 40% of dasyurids travelled above the floor (e.g. on raised shelves) at the Compton Road fauna underpasses (see Box 8.2) (Bond and Jones 2008). Similarly, Woltz *et al.* (2008) found that North American green frogs (*Rana clamitans*) preferred soil-lined and gravel-lined tunnels to concrete and PVC-lined tunnels.

Placement

Fauna underpasses are likely to be most effective in reducing road-kill in areas where fauna–vehicle collisions are most likely to occur (i.e. hotspots). To improve their effectiveness, underpasses should be placed in an area of suitable habitat and within the natural paths of target species. The surrounding land use should be considered when planning a land bridge. Ideally, the habitat to be connected by the fauna underpass should have a reasonable prospect of remaining suitable habitat in the future.

Maintenance

Fauna underpasses should be maintained to help ensure that they function as intended. Maintenance includes ensuring that silt or debris from flooding does not build up in culverts to levels that inhibit wildlife passage. If fauna exclusion fencing is used in conjunction with the

Table 8.4. Some design factors that influence the effectiveness of a fauna underpass

Factor	Details
Fauna exclusion fencing	Fencing helps to prevent animals from entering the roadway, as well as to funnel individuals into the fauna underpass.
Dimension	Species may prefer particular dimensions (e.g. long and narrow or short and open).
Substrate	Inclusion of a natural substrate (placing rocks and gravel on underpass floor) may encourage some animals to use the underpass. Moisture is required for certain species (e.g. amphibians).
Furniture	Hollow logs and pipes placed within underpasses can encourage use by some mammal species (e.g. walkways; see Figure 8.10).
Disturbance by people	Human access of underpasses should be restricted because the use of crossing structures can be influenced by noise and human activity.
Approaches	Some species may prefer tunnels that have a clear view to the other end whereas other species may prefer cover.

underpass, it must be maintained to ensure animals cannot pass through (e.g. no tears or gaps in or under fence and no vegetation growing over it).

Impacts on other aspects of the environment

It has been suggested that predators can use fauna underpasses to aid in capturing prey (Fahrig et al. 1995; Norman et al. 1998). Predators may potentially learn that underpasses are advantageous places to hunt because prey animals within the underpass have limited means of escape. However, evidence that underpasses function as prey-traps is largely anecdotal (Little et al. 2002). Nonetheless, instances of decline in a species after installation of a fauna underpass should not be ignored. For example, a severe decline in a local population of southern brown bandicoots (*Isoodon obesulus fusciventer*) coincided with an underpass observed to be used by red foxes (*Vulpes vulpes*) in Perth (Harris et al. 2010).

The potential for increased predation may be reduced by providing moderate vegetative cover at the entrances of the fauna underpasses as well as raised shelves, hollow logs and pipes within the underpass (Norman et al. 1998; Bond and Jones 2008). In other cases, control of introduced predators is necessary (e.g. active removal by trapping or by removing suitable habitat).

Canopy bridges

Overview	
What:	A suspended structure intended to facilitate the movement of scansorial (i.e. climbing) fauna over developments.
Where:	Beneficial when a development (e.g. a road) or a clearing needs to be spanned to provide continuity of habitat or facilitate movement.
Pros:	Links fauna habitat and facilitates movement of scansorial fauna.
Cons:	Predators may potentially use canopy bridges as prey traps.
Cost:	Low (e.g. ropeway)–medium (e.g. rope tunnel).
Companion measures:	Natural habitat linkages, land bridges and glide poles (all this chapter).
Alternative measures:	Avoid impacts to fauna movement (Chapter 4); natural habitat linkages and glide poles (this chapter).

Figure 8.14. Common ringtail possums (*Pseudocheirus peregrinus*) with young on a rope bridge (Photo: R van der Ree)

Description and application

A canopy bridge is a suspended structure (e.g. rope bridge) intended to facilitate the movement of arboreal and scansorial (i.e. climbing) fauna (e.g. possums) in areas where the tree canopy is no longer continuous (Figure 8.14). Canopy bridges could be used to retain habitat connectivity for different types of developments that lead to the disruption of the tree canopy, such as roads, railway lines and utility easements. Canopy bridges can be used as a permanent measure where maintenance of an intact canopy is not feasible or as a temporary measure where a continuous tree canopy is gradually reforming (Figure 8.15).

Until recently, arboreal mammals were largely overlooked by road authorities attempting to mitigate the impacts of roads on fauna (Taylor and Goldingay 2009). Canopy bridges are an important step forward in reconnecting habitats because many predominantly and obligatory arboreal mammals may not make use of land bridges, fauna underpasses or glide poles (as discussed in this chapter).

It is not uncommon for animals to use pre-existing structures built for other purposes to cross developments. Common ringtail possums (*Pseudocheirus peregrinus*) are often observed using powerlines strung across roads to move around in urban areas.

Types of canopy bridges include 'ropeways' (a single strand rope), 'pole bridges' (bridge made from wooden poles), 'rope ladders' (also known as possum and monkey bridges) and 'rope tunnels' (Figure 8.16). Canopy bridges are made from marine-grade rope strung from trees or artificial structures, such as vertical poles with concrete footings. Across roads, canopy bridges are strung approximately 7 to 12 m high, often with steel cables to reduce lag in the rope. Ideally, each side of the rope walkway is attached to established trees within suitable habitat to enable fauna to enter and exit the bridge. Multiple ropes can be strung to multiple trees at each end to improve entry and exit.

Figure 8.15.   A rope ladder canopy bridge connecting remnant vegetation. Saplings have been planted beneath the canopy bridge, which, when mature, will provide long-term connectivity in place of the canopy bridge (Photo: R Whitney)

Weston (2003) provided a comprehensive overview of the use of the different types of canopy bridges in Australia and around the world. For example, ropeways have been used overseas in the United States of America and Europe to promote safe movement, mainly for squirrels (*Sciurus* spp.). Pole bridges have been used in Brazil to connect habitat and reduce road mortality of the endangered black lion tamarin (*Leontopithecus chrysopygus*) (Valladares-Padua *et al.* 1995). Ladder bridges have been erected for the use of monkeys in a number of countries including Taiwan, Belize, Mexico and Kenya (see Weston 2003).

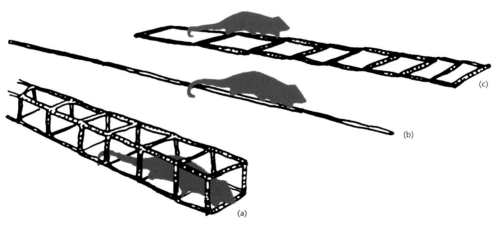

Figure 8.16.   Different types of canopy bridges: (a) rope tunnels, (b) ropeways or pole bridges and (c) rope ladders

Box 8.4. Canopy bridges provide connectivity for rainforest fauna

The use of rope tunnel and rope ladder canopy bridges by fauna within the rainforests of the Wet Tropics World Heritage Area, north-east Queensland was monitored using remote photography, spotlighting and scat and hair identification in the early 2000s by Goosem *et al.* (2005). The rope tunnel and rope ladder canopy bridges were installed over roads carrying relatively low traffic volumes of between 4 and 150 vehicles per day. Both types of canopy bridge were approximately 14 m long and the sites chosen had no natural canopy connections across the roads.

Three rare rainforest ringtail possums – the lemuroid ringtail possum (*Hemibelideus lemuroides*), Herbert River ringtail possum (*Pseudochirulus herbertensis*) and green ringtail possum (*Pseudochirops archeri*) – were specifically targeted during the study (Goosem *et al.* 2005). The lemuroid ringtail possum is an obligate arboreal species (i.e. never descends to ground level) (Winter *et al.* 2008), while the Herbert River ringtail possum will descend to the ground (Winter and Moore 2008).

All three of the target species used the rope tunnel canopy bridge, while two of the target species – the lemuroid ringtail possum and Herbert River ringtail possum – were observed using the rope ladder canopy bridge (Goosem *et al.* 2005). Other canopy bridge designs (e.g. a wider rope ladder and single strand of rope) were tested, but the results were inconclusive as to their effectiveness.

Within Australia, canopy bridges have been installed in at least nine locations and all of these were across roads (van der Ree *et al.* 2008b). Locations within Australia include over the Pacific Highway north of Newcastle in New South Wales (Bax 2006), within the rainforests of the Wet Tropics World Heritage Area, north-east Queensland (Box 8.4) and beside a land bridge at Hamilton Road, Brisbane (Figure 8.1).

Effectiveness

Three types of canopy bridges – ropeways, rope ladders and rope tunnels – have undergone limited trials within Australia to date (Goosem *et al.* 2005; Bax 2006; Veage and Jones 2007; van der Ree *et al.* 2008b). At least 11 Australian species have been observed using these canopy bridges as crossing structures (Table 8.5), with many birds also using the structures as roosting perches.

It is not known to what extent canopy bridges would be used by gliding mammals. Squirrel gliders have been observed using rope tunnel canopy bridges (Bax 2006) and rope ladder canopy bridges (van der Ree *et al.* 2008b). However, glide poles are an alternate wildlife crossing structure that are also being trialled to assist movement of gliding mammals (see next section).

Previous studies have found that it can take some time before a canopy bridge is used by fauna. Goosem *et al.* (2005) observed that more than 5 months passed before a rope ladder canopy bridge was used, after which crossing rates then became common, rising to around one crossing every hour. However, this delay in usage is not always the case. Overseas, Valladares-Padua *et al.* (1995) found that black lion tamarins (*Leontopithecus chrysopygus*) and capuchins (*Cebus apella*) began using a pole bridge as soon as it was installed.

Not enough detailed studies have been undertaken to determine the effectiveness of canopy bridges against the criteria of Forman *et al.* (2003) (see introduction to this chapter) and in

Table 8.5. Australian species observed using canopy bridges

Species	Type of canopy bridge	Reference
Common brushtail possum (*Trichosurus vulpecula*)	Rope tunnel	Bax 2006
	Rope ladder	van der Ree *et al.* 2008b
Common ringtail possum (*Pseudocheirus peregrinus*)	Rope ladder	Veage and Jones 2007; van der Ree *et al.* 2008b
Squirrel glider (*Petaurus norfolcensis*)	Rope tunnel	Bax 2006
	Rope ladder	van der Ree *et al.* 2008b
Lemuroid ringtail possum (*Hemibelideus lemuroides*)	Rope tunnel, rope ladder, ropeway*	Goosem *et al.* 2005
Herbert River ringtail possum (*Pseudochirulus herbertensis*)	Rope tunnel, rope ladder, ropeway*	
Green ringtail possum (*Pseudochirops archeri*)	Rope tunnel, rope ladder**	
Fawn-footed melomys (*Melomys ceruinipes*)	Rope tunnel, rope ladder**	
Coppery brushtail possum (*Trichosurus johnstonii*)	Ropeway*, rope ladder	
Striped possum (*Dactylopsila trivirgata*)	Rope tunnel, rope ladder	
Long-tailed pygmy possum (*Cercartetus caufatus*)	Rope ladder**	
Lumholtz's tree-kangaroo (*Dendrolagus lumholtzi*)	Rope ladder	

*Use was inconclusive owing to no direct or photographic evidence.
**Use was inconclusive because canopy connections were present above the canopy bridge.

maintaining or improving the viability of local populations. Factors likely to influence the effectiveness of canopy bridges include its design, placement and maintenance.

Design

The key elements for the design of effective canopy bridges for Australian fauna are not well understood, because their use is still in its infancy. Different arboreal species probably have preferences for particular types of canopy bridges so, in the absence of relevant studies, various designs may need to be trialled. Disturbance from vehicle headlight glare has been suggested to influence the effectiveness of canopy bridges (e.g. Bax 2006).

Placement

The optimal placement of a canopy bridge depends on the home range requirement of the target species. For the lemuroid ringtail possum (*Hemibelideus lemuroides*), Wilson *et al.* (2007) recommended providing one canopy linkage for every home range – in this case, every 100 m – to maintain social integration and genetic diversity across a linear barrier. Weston (2003) recommended that rope ladders should be installed every 100–120 m, at least 6 m above the road surface, in areas of the Wet Tropics of north Queensland where there has been a loss of canopy connectivity.

To improve the effectiveness of the structure, it should be placed in an area of suitable habitat and within the natural paths of target species. Obviously, if there is already sufficient canopy connection a canopy bridge will be less likely to be used.

Maintenance

Maintaining rope integrity for the life of the rope bridge is important to make sure that it does not fall across the road. This is especially important from a road safety perspective. Rope may need to be replaced periodically, depending on environmental conditions. If marine-grade rope is used, this probably would be required less frequently. Experiments trialling metal chain or steel cables may be an option, instead of using rope.

Impacts on other aspects of the environment

Similar to land bridges and underpasses (Fahrig *et al.* 1995), it is a possibility that predators might use canopy bridges as prey traps. Whether this happens or not is not known. The rope tunnel design was thought to provide protection from aerial predators because animals can travel within the tunnel structure (see Weston 2003). However, several investigators have observed that most animals using the rope tunnel design cross along the top surface of the structure, not inside it (Goosem *et al.* 2005; Bax 2006).

Glide poles

Overview	
What:	Tall poles fitted with crossbars at the top that are intended to provide movement pathways for gliding fauna.
Where:	Situations where habitat patches are separated by a distance greater than the gliding capabilities of the species concerned.
Pros:	Links fauna habitat; facilitates movement of gliding fauna.
Cons:	Gliding fauna may be more vulnerable to predation while on glide poles. In some instances, glide poles placed beside roads or in the centre median strip may contribute to increased mortality if assisting gliders to cross a road increases their chances of colliding with vehicles.
Cost:	Low
Companion measures:	Natural habitat linkages, land bridges and canopy bridges (all this chapter).
Alternative measures:	Avoid impacts to fauna movement (Chapter 4); natural habitat linkages and canopy bridges (this chapter).

Description and application

Glide poles (or often called 'glider poles') are tall poles fitted with one or two crossbars (launch beam) at the top to provide movement pathways for gliding fauna, such as squirrel gliders (*Petaurus norfolcensis*) in situations where natural habitat patches are separated by a distance greater than the gliding capabilities of the species concerned. Retaining contiguous habitat patches is desirable to facilitate movement (e.g. for dispersal or foraging). Glide poles are a relatively new type of artificial wildlife crossing structure and are being pioneered in Australia. Glide poles are made from vertical tree logs (e.g. wooden power poles) with horizontal crossbars to provide similar launch and landing opportunities as a tree trunk (Figure 8.17).

Glide poles could be used to retain or re-establish habitat connectivity for different types of developments that involve the removal of trees used by gliding fauna. In particular, glide poles show promise for linking habitat across linear infrastructure such as roads, railway lines and utility easements by facilitating movement of gliding fauna above infrastructure (or below in the case of powerlines, if safe to do so).

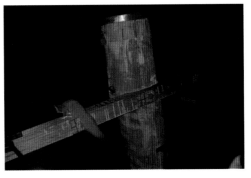

Figure 8.17.   (a) Glide poles installed on a land bridge over Compton Road, southern Brisbane (Photo: D Gleeson) and (b) a squirrel glider (*Petaurus norfolcensis*) on a glide pole at Warrenbayne Road, Victoria (Photo: R van der Ree)

Planting suitable trees in preference to erecting a series of glide poles is usually a cheaper and longer term solution because this typically requires less maintenance. However, in many cases, planting trees is not possible or it may take too long (10 years or more) for planted trees to reach a height suitable for facilitating routine gliding (Ball and Goldingay 2008). Glide poles could be used permanently when it is not possible to bridge a gap in the canopy with trees, or temporarily until trees are tall enough to be used by gliding fauna in order to avoid negative consequences (including the possibility of extinction) to a local glider population (Box 8.5). Glide poles could also be erected on land bridges (Figure 8.17a), particularly if the soil is too shallow to support the growth of large trees suitable for gliding. Glide poles can be erected on the road verge as well as in the centre median strip (Goldingay and Taylor 2009; Goldingay *et al.* 2011).

Box 8.5.   Evidence that the squirrel glider can use glide poles

The first peer-reviewed published study in the world to use timber poles to successfully provide habitat connectivity for a gliding mammal species was undertaken in 2005 by Ball and Goldingay (2008) at Padaminka Nature Refuge near Mackay in north Queensland. The study used five untreated hardwood power poles installed 16–22 m apart (producing pole-to-pole distances of 16–40 m) between two remnant patches to test whether squirrel gliders (*Petaurus norfolcensis*) could traverse between them successfully.

The poles installed by Ball and Goldingay (2008) had a diameter of 30 cm. They were buried 1.5 m into the ground and stood 12 m high once installed. A crossbar was attached to the top and it was later found that most individuals used the crossbar to launch into a glide.

Ball and Goldingay (2008) tested the usage of the glide poles by releasing 22 individuals onto the poles at night and observing their behaviour. Two individuals were radio-tracked and hair-sampling devices were fitted to the poles. The researchers demonstrated that the squirrel gliders were using the poles to traverse open areas between habitat patches. They found that individual squirrel gliders were only captured in both remnants after installation of the glide poles and there was evidence of ongoing use of the poles by squirrel gliders over a period of 12 months.

Effectiveness

Glide poles have undergone limited trials to date. The first experimental glide poles were erected at Nowra, on the south coast of New South Wales in 1995 (Goldingay *et al.* 2011). These poles were installed either side of a powerline easement, but it was never established whether gliders used the poles.

Glide poles have since been installed at other locations within Australia including along a land bridge at Compton Road, southern Brisbane (Bond and Jones 2008; Goldingay *et al.* 2011) (Figure 8.17a), on a land bridge over Hamilton Road, Brisbane and at Padaminka Nature Refuge near Mackay in north Queensland (Ball and Goldingay 2008) (Box 8.5). Glide poles at these locations are known to have been used by gliders (Goldingay *et al.* 2011).

The effectiveness of glide poles is not well understood, because these wildlife crossing structures are still undergoing trials. Although there is evidence of gliders using glide poles installed on land bridges spanning roads with heavy traffic flows, it is not yet known whether gliders will glide directly over very busy roads using glide poles installed at the sides and within the median strip (Goldingay *et al.* 2011).

It has been confirmed that squirrel gliders have the capability to use appropriately designed and spaced glide poles (Ball and Goldingay 2008; Goldingay *et al.* 2011). However, information regarding usage of glide poles is not yet known for other gliding species (Taylor and Goldingay 2009). Mahogany gliders (*Petaurus gracilis*) have been observed using wooden power poles to launch into a glide, suggesting that this species may use purpose-built glide poles (Asari *et al.* 2010).

Although it is not yet known whether the greater glider (*Petauroides volans*) will use glide poles, Taylor and Goldingay (2009) used population viability modelling to investigate whether glide poles would reduce the probability of extinction of a smaller sub-population of greater glider, assuming that they would use them. They demonstrated that even a relatively low rate of dispersal across a major road (Compton Road, Brisbane) facilitated by glide poles would reduce the probability of extinction of a smaller sub-population.

Design

The physical capabilities and behavioural tendencies of the target species will need to be taken into consideration when deciding on the design and spacing of glide poles. The optimal height and spacing of glide poles would depend on the glide distance capability for the target species. Glide distance is directly proportional to the launch height (Ball and Goldingay 2008; Goldingay and Taylor 2009) (Figure 8.18). Ball and Goldingay (2008) found that the maximum successful glide distance between 12 m high poles for squirrel gliders was 25 m and those individuals that attempted to glide 30 m fell short. Goldingay and Taylor (2009) studied the gliding performance of squirrel gliders in forest remnants in South-East Queensland and predicted that 12 m high glide poles would be adequate beside two-lane roads, but additional poles in the centre median strip would be required for four-lane roads. Trials would be needed to determine the gliding performance of other gliding species.

The optimal design of the launch beam is not yet known. A variation of the design described here is a launch beam that extends towards the gap in the canopy, as installed by Wagga Wagga City Council for squirrel gliders (Goldingay *et al.* 2011).

Placement

To improve the effectiveness of glide poles, they should be placed in areas linking suitable habitat of the target species. Whether glide poles can mitigate the impact of a development would depend on constraints present at each situation. For example, in some locations, road

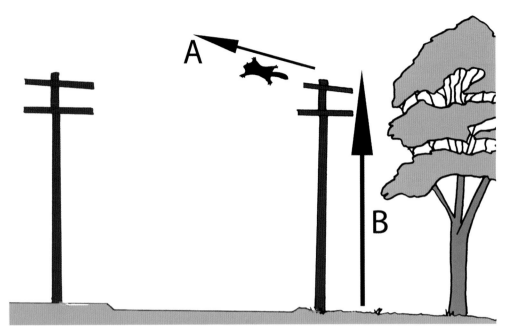

Figure 8.18.   Glide distance (A) is directly proportional to the launch height (B)

width or design make it impossible to incorporate glide poles at appropriate distances for glider use. Goldingay *et al.* (2011) concluded that glide poles can enable squirrel gliders to cross large gaps in the canopy of around 60–70 m.

Impacts on other aspects of the environment

A small area of ground disturbance is required for the physical placement of glide poles. There is normally some flexibility where they are placed, so particular habitat features or significant vegetation can usually be avoided.

The cavities and foliage of trees normally provide gliding fauna with protection from predators (e.g. owls). Gliding fauna may be more vulnerable to predation while on glide poles. Attaching PVC pipe to glide poles at various heights has been used as a way of providing substitute refuge areas (e.g. Ball and Goldingay 2008), but no published studies have yet shown whether these are effective. Aluminium shields to protect gliders from avian predators hunting from above have been added to glide poles recently erected as part of the Hume Highway duplication in south-western New South Wales (van der Ree *et al.* 2008).

In some instances, glide poles placed beside roads or in the centre median strip may contribute to increased mortality if assisting gliders to cross a road puts them in the path of oncoming vehicles, but there is no evidence of this at present (McCall *et al.* 2010; van der Ree *et al.* 2010).This would be especially true if gliders that attempt to cross a road do not have suitable landing opportunities or if the poles are not of adequate height for the gliders to remain above the vehicles using the road (including trucks). Although glide poles have been shown to be used by gliding fauna, thereby increasing habitat connectivity, the possibility that there could be some mortality arising from their use should not be ignored (McCall *et al.* 2010), and is a justification for further research on this innovative management tool (Goldingay *et al.* 2011).

Escape routes

Overview	
What:	Provides fauna with a means of escaping a threatening situation.
Where:	Types of developments that can trap fauna resulting in injury or death or otherwise threaten fauna (e.g. embankments along roads that animals cannot climb to avoid an oncoming vehicle).
Pros:	Prevents animal fatalities or injury.
Cons:	Animals may not locate the escape route.
Cost:	Low (e.g. length of rope)–medium (earthen ramp).
Companion measures:	Fauna exclusion fencing (Chapter 7); natural habitat linkages (this chapter).
Alternative measures:	Avoid threatening situations (Chapter 4) or by physically keeping fauna away from hazards, such as using coverings and fauna exclusion fences (Chapter 7).

Description and application

Escape routes (e.g. climbing aids and earthen ramps) provide fauna with a means of escaping a threatening situation. Some types of development can trap fauna, which can result in injury or death. An animal that inadvertently falls into a steep-sided water body, for example, will eventually drown if it cannot escape. Escape routes should be considered when it is not possible to avoid a threatening situation by re-designing a component of the development (e.g. avoiding the creation of embankments along roads) or by physically preventing fauna from reaching the hazard, such as by erecting fences or using coverings (Chapter 7).

Many different types of wildlife escape routes have been created to suit a wide variety of developments. Examples of escape routes are provided in Table 8.6. They can usually be incorporated into the initial design of a development or retrofitted to an existing development.

Climbing aids

Wildlife can become trapped when they are unable to climb over or out of a structure. Climbing aids could be used down holes with or without water (e.g. pits, trenches, dams with steep sides and residential pools), residential fencing and road safety barriers.

Table 8.6. Examples of escape routes

Type	Description
Climbing aids	Materials (e.g. rope, mesh and wooden poles/logs) that can be used by climbing animals to escape from a hole or to enable fauna to cross a barrier.
Refuge poles	Modified poles installed to provide fauna with refuge from predators. Also called 'shelter poles'.
Water body escape ramps	Ramps installed in steep-sided water bodies as a means for animals to escape and so avoid potential drowning.
Earthen ramps	Ramps of mounded earth that can be used by animals to access an elevated area (e.g. along roads flanked by steep embankments and/or sheer drop-offs). One-way ramps can be installed in conjunction with a fauna exclusion fencing so trapped animals are able to escape from a roadway.
One-way gates	One-way gates can be incorporated into fauna exclusion fencing to allow trapped animals to exit a roadway.

Figure 8.19. Escape pole that allows climbing fauna (e.g. koalas) to climb over a fauna exclusion fence if trapped on the wrong side (Photo: J Gleeson)

A length of rope or netting secured to a fixed point above ground level, for example, can promote the egress of fauna from holes or bodies of water. A float secured to the end of the rope may be required if used in water. Logs can be secured against residential fences to provide a means for animals to climb to safety if confronted by domestic dogs. Similarly, escape poles installed up against fauna exclusion fencing can be used by some trapped animals to escape from a roadway (Figure 8.19). These animals can otherwise become trapped on the roadway where they are vulnerable to being hit by vehicles if they are unable to find their way back through, over or around the fence. Netting and/or wooden poles have been used to provide a means for koalas (*Phascolarctos cinereus*) to climb over barriers (Figure 8.19).

Refuge poles

Refuge poles are effectively artificial trees that are aimed at providing arboreal animals (e.g. koalas) with a place to hide from predators in open areas (Figure 8.20). Arboreal animals attempting to cross open areas, cleared of vegetation, are at increased risk of predation. It takes many years for planted trees to reach a height suitable for refuge and sometimes it is not possible to plant trees at all. In these situations, refuge poles could provide an alternative. Shade cloth may be fixed to the top of the refuge pole to provide shade for arboreal animals that need to take refuge during the day, as well as protection from birds of prey. Multiple refuge poles can be installed in open areas between two habitat patches. If refuge poles are to be installed, care should be taken so they are positioned to maximise their usefulness (i.e. not close to existing trees that are likely to be preferentially used as refuge).

Water body escape ramps

Animals that fall into a steep-sided water body (e.g. water management ponds used in association with mining, livestock water troughs or swimming pools) will eventually drown if unable

Figure 8.20.   A refuge pole can be climbed by a koala (Photo: D Gleeson)

to escape. Water body escape ramps can be installed to prevent this from happening. Such escape ramps should be made from easily gripped materials (e.g. concrete with horizontal etchings or metal grating) and will be of maximum use if angled gently. Escape ramps should be located at the perimeter of the water body so that an animal trying to escape will run into it as it moves around the edges of the water body, trying to get out. Ramps made from polyethylene and designed to suit use in residential swimming pools have recently become commercially available.

Earthen ramps

A gently sloped earthen ramp can enable fauna to access a higher ground level (Figure 8.21). Some roads, particularly in mountainous areas, are flanked by steep embankments and deep gutters. This makes it difficult for animals already on these roads to flee oncoming vehicles. In these situations, animals tend to respond to an approaching vehicle by running down the middle of the road until flat ground unobstructed by dense vegetation is available for its escape (Magnus *et al.* 2004). In these situations, earthen ramps may provide animals with a means of escaping traffic. Earthen ramps could also be used to enable fauna to escape trenches.

One-way ramps have been installed to assist the escape of fauna trapped in a fenced roadway (Figure 8.22). The ramp leads to a gap in the fence that animals can pass through. Animals already on the other side of the fence (i.e. outside of the roadway) cannot gain access to the road due to the high wall of the ramp terminus, which prevents animals climbing up through the gap in the fence. The optimal height of the ramp terminus would vary depending on the species. Care should be taken to ensure that the height is not so great that animals exiting the roadway are injured when they attempt to move down off the ramp terminus. A similar type of ramp was incorporated into the fauna exclusion fencing along the Tugun Bypass on the Queensland/New South Wales border (Queensland Department of Transport and Main Roads 2010).

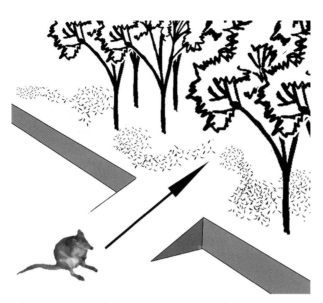

Figure 8.21.   Animals can use an earthen ramp to escape to higher ground

Swinging gates

Swinging one-way gates can be incorporated into fencing to allow trapped animals to exit a fenced area in one direction. Swinging gates are also used to minimise fence damage by animals trying to break through a fence. For example, the installation of low flaps (called 'wombat gates') can reduce the fence damage caused by wombats in agricultural settings (Statham 2009; Triggs 2009). Swinging gates could be incorporated into particular sections of a fence for specific fauna where it is desirable for the animals to move from one side of the fence to another. Also refer to fauna inclusion fences described in the next section.

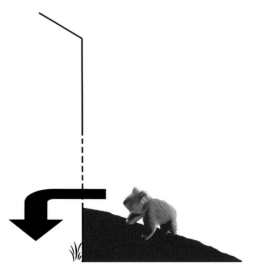

Figure 8.22.   A cross section of a one-way ramp for animals. The ramp leads to a gap in the fence that animals can pass through. The high wall of the ramp terminus prevents animals from moving back through the gap

Effectiveness

There is a lack of published research on the effectiveness of escape routes. All escape routes described in this chapter have been applied in an attempt to reduce impacts on wildlife. The effectiveness of a particular escape route will depend on a variety of factors, including the development, the threat and the behaviour and physical capabilities of the species that are likely to use it.

In general, water body escape ramps are effective for providing a range of fauna escape routes from water bodies. Chapter 10 discusses how farm dams can be made more amenable to wildlife by adopting a gradual bank to the dam. Similarly, practical experience has shown that climbing aids are effective for animals that are able to climb. Road safety barriers modified with, firstly, plastic mesh and, secondly, horizontal wooden poles placed no more than 50 cm apart were both successfully climbed over by some koalas during trials (Queensland Department of Transport and Main Roads 2010).

There is a general lack of documented evidence of the effectiveness of refuge poles. These refuges are being trialled in northern Queensland for the tree-kangaroo (*Dendrolagus* spp.). They could be used experimentally for other species.

Design

The effectiveness of earthen ramps (including one-way ramps) and one-way gates as escape strategies depends on effective design. Various designs of each have been trialled as described below.

Earthen ramps provide an effective solid platform for animals to exit a threatening situation, depending on the compaction of the soil and hydrology of the area. Applying bitumen or concrete coatings may improve their integrity. Earthen ramp escape routes were part of a variety of techniques that were used in an attempt to reduce the impact of road traffic on the eastern quoll and Tasmanian devil along a tourist road into the northern end of the Cradle Mountain-Lake St Clair National Park in Tasmania (Jones 2000). Earthen ramps were constructed at around 25 m intervals in areas where escape routes were deemed necessary because deep gutters and steep embankments were likely to trap fauna on the road. The ramps bridged the road verge to the top of the embankment and shelter for fauna was provided under the ramp in the form of 2 m lengths of drainage pipe running parallel to the road. The earthen ramps and pipes were used by Tasmanian devils and wombats, while macropods used the ramps. The combined techniques appeared to have successfully reduced the impact of traffic on the eastern quoll and Tasmanian devil, but it was not possible to determine how much, if any, of this was due to the escape routes (Jones 2000).

One-way ramps will only suit animals that are capable of jumping the distance from the top of the ramp to the natural ground level on the other side of the fence. If the distance is too great for the target species, an alternative design can be used incorporating a more suitable distance. Research to determine the effectiveness of escape routes over, or through, roadside fencing for deer in the United States of America revealed that one-way earthen ramps were eight to eleven times more effective than one-way steel gates (Bissonette and Hammer 2000).

One-way gates should be large enough to accommodate the largest animal that is likely to use them. Gates may be shunned by species that are hesitant to squeeze through the gate (Bissonette and Hammer 2000). An alternative for these species is installation of a one-way ramp that resembles natural topography. One-way ramps can be made more attractive to animals by encouraging native plants (e.g. grasses) to grow on them.

Placement

The spacing of escape routes should be site-specific, but, in general, the more escape routes available lessens the risk to any trapped animals. As an example, in the United States of

America, Bissonette and Hammer (2000) recommend one-way earthen ramps should be placed at around 0.5 km intervals for use by deer.

Maintenance

Ongoing maintenance is required for the medium- and long-term effectiveness of escape routes. Material used in climbing aids and refuge poles may need to be replaced periodically. Earthen ramps are likely to require frequent inspections and maintenance, due to erosion.

Impacts on other aspects of the environment

The impacts on other aspects of the environment are likely to be minimal, if any. A small area of ground disturbance may be required for the physical placement of refuge poles or ramps. There is normally some flexibility where they are placed, so particular habitat features or significant vegetation can usually be avoided.

Water body escape ramps generally do not require land clearance because they can often be located within the overall pond structure. Due to the slope of these ramps, they are likely to increase fauna visitation to the water body, which may or may not be desirable. If this is not desirable perhaps fencing the water body could be considered.

Fauna inclusion fences

Overview	
What:	Fences that allow animals to move freely from one side of the fence to the other.
Where:	Useful where development requires a fence (e.g. to demarcate property boundaries), but native animals do not need to be excluded from the development site.
Pros:	Improves opportunities for fauna to access habitat resources (e.g. food, shelter or nesting sites) and other individuals in a population.
Cons:	Can impede movement of some, usually larger, species.
Cost:	Low–medium, depending on the distance to be fenced.
Companion measures:	Natural habitat linkage (Chapter 8).
Alternative measures:	Modify the development to avoid using a fence or place the fence where it is unlikely to obstruct the movement of fauna (Chapter 4); escape routes (this chapter).

Description and potential applications

Fauna inclusion fences are designed to allow animals, generally smaller animals, to move freely from one side of the fence to the other. This improves their opportunities to access habitat resources (e.g. food, shelter or nesting sites) and other individuals in a population. Fauna inclusion fences can be designed to enable fauna to pass through, over or under the fence.

These fences are usually applied in situations where human access is to be restricted (e.g. property boundaries), but where native animals do not need to be excluded from the development site (Figure 8.23). Fauna inclusion fences are increasingly being incorporated into housing developments where domestic animals are not permitted, such as the Koala Beach Housing Project, north of Pottsville, northern New South Wales coast (Beatly and Newman 2009). Fauna inclusion fences are less likely to sustain damage by animals trying to break their way through than a standard fence.

Figure 8.23. Fauna inclusion fence at Deagon Wetlands, Brisbane (Photo: D Gleeson). Note the lower half of the fence is left open for smaller animals to pass under

Although fauna inclusion fences are designed to enable fauna movement, they may hinder the free movement of some, possibly larger, animals (e.g. kangaroos or emus). It would be best to design developments to avoid installing a fence or avoid installing it where it is likely to obstruct the movement of fauna (Chapter 4).

Before replacing an existing standard fence with a fauna inclusion fence, it should be considered whether the existing fence is protecting fauna from an existing threat (e.g. road or domestic animals). For example, if a semi-rural property was bordered by native bushland on one side and a busy road on the other, it may be better to use a fauna inclusion fence adjacent to the bushland and a standard fence or fauna exclusion fence on the other side, adjacent to the road.

The optimal design of a fauna inclusion fence will depend on the traits of the animals that will pass through, over or under it. Examples of animals grouped according to their characteristics and design that best suits them are provided in Table 8.7. Fauna inclusion fences work on the reverse concepts of fauna exclusion fences discussed in Chapter 7.

Table 8.7. Examples of animal characteristics and fence design elements that best suits them

Type of animal	Design of fence
Climbers (e.g. possums and koalas)	Chain wire provides a medium that can be climbed by most agile climbing animals. To enable fauna to crawl along (e.g. possums), the top of the fence should be made from a ridged material.
Diggers (e.g. wombats and echidnas)	Space left between the ground and fence mesh for them to pass under.
Jumpers (e.g. kangaroos, wallabies and bandicoots)	Height of the fence to suit jumping over, or space left between the ground and fence mesh for them to pass under.
Small size (e.g. amphibians and reptiles including turtles)	Space left between the ground and fence mesh for them to pass under.

Figure 8.24. Example of a wooden property fence that allows most fauna to pass from one side to the other (Photo: J Gleeson)

Fauna inclusion fences can be made from many different materials (e.g. wood or chain wire) (Figures 8.23 and 8.24). Common design features include: low overall height of the fence (approximately 1200 mm), the bottom of the fence offset at least 300 mm above the ground and the top of the fence allows climbing animals to freely pass.

Many existing fences can be modified to improve fauna movement. Solid fences made from timber or concrete can be modified by installing escape poles (described in this chapter). Various alternatives to the use of barbed wire, some of which also aid fauna movement (e.g. virtual fence technology), are described in the next section. As described earlier, swinging gates incorporated into a standard fence can facilitate movement of fauna past the fence at a chosen location and also reduce fence damage. For example, the installation of gates for wombats can reduce fence damage they would otherwise cause (Statham 2009; Triggs 2009).

Effectiveness

There is a lack of published research of the effectiveness of fauna inclusion fences. Practical application has shown that fauna exclusion fences improve animal movement. In contrast to fauna exclusion fencing (Chapter 7), vegetation can be permitted to grow near or over the fence because it does not matter if its substrate assists animals climb over the fence.

Impacts on other aspects of the environment

As mentioned above, fauna inclusion fences may still hinder the free movement of some animals, so the fences can still act to guide fauna in a particular direction and it is important that they do not direct fauna towards a threat (e.g. onto a busy road).

Alternatives to barbed wire

Overview	
What:	Other ways of maintaining the functioning of fences as a barrier to stock and other animals that have traditionally incorporated barbed wire in their design.
Where:	Useful where developments (e.g. farms) have used, or are considering using, barbed wire.
Pros:	Reduces the risk of wildlife fatalities and injuries.
Cons:	Alternative fences may still impede fauna movement.
Cost:	Low–medium.
Companion measures:	Reducing attractive habitat (Chapter 7); Escape routes (this chapter).
Alternative measures:	Design modifications to avoid using a fence (Chapter 4); fauna exclusion fences (Chapter 7); fauna inclusion fence (this chapter).

Description and potential applications

Barbed wire is still widely used in Australia and throughout the world for various reasons, but predominantly in the agricultural industry to retain livestock (particularly cattle) within paddocks.

Barbed wire fences can be a significant cause of fatalities in birds, arboreal mammals, ground mammals, reptiles and bats (Allen and Ramirez 1990; van der Ree 1999; Booth 2007a, 2007b; Ley and Tynan 2008). Fauna that become entangled can be injured or experience a slow and painful death. The Wildlife Friendly Fencing Project, funded by the World Wide Fund for Nature (WWF), has helped to raise awareness of the issue in Australia (Tolga Bat Hospital 2010).

Not all barbed wire fences present the same risks to local fauna. The risk can be greater for newly installed fences, fences that run across fauna movement paths, such as across farm dams (Figure 8.25), other water bodies or along ridges (Booth 2007a) and those in fragmented habitat (van der Ree 1999). Barbed wire fences also pose a greater risk when used within the range of a densely populated species that are susceptible to entanglement (van der Ree 1999).

Over time, landholders have sought alternatives to using barbed wire, and not only because of the threat they pose to native animals. Barbed wire is relatively costly, cumbersome to install, has a relatively short life span, poses a risk of damaging or injuring livestock and requires frequent maintenance due to sagging wires (Kubik 2007). Fencing is used not only in agriculture, but also for conservation purposes as part of revegetation and restoration programs.

It would be best to investigate alternatives to avoid installing a fence altogether, or avoid installing it where it is likely to obstruct the movement of fauna. van der Ree (1999) proposed that constructing fences diagonally across the corner of a paddock (i.e. cutting the corner rather than fencing to the intersection point) would minimise fence length and minimise impact on fauna attempting to cross two fence lines that intersect at a corner.

Readily available alternatives to barbed wire fences and ways to modify existing barbed wire fences are provided in Table 8.8. In addition to these measures, other innovative alternatives, such as virtual fences (Box 8.6), are also being researched.

Effectiveness

There is a dearth of research into the effectiveness of alternatives to barbed wire fences. Alternatives must successfully contain livestock as well as reduce the risk to animals in order to be effective. Landholders want cost-effective alternatives that are easy to install and are long lasting.

Figure 8.25.   A barbed wire fence that poses a high risk to local fauna (Photo: J Gleeson)

Fencing with high-tensioned plain wire was first developed in New Zealand in the 1970s (Kubik 2007) and has since been successfully used to contain sheep and similar livestock. Fences made from high-tensioned plain wire are in many ways superior to barbed wire fences because they are comparatively cheaper, easier to install, reduce the risk of injury to livestock and require less maintenance (Kubik 2007; Booth 2007b).

Farmers in Australia have raised concerns that a high-tensioned plain wire fence is not effective for cattle because they can push through the fence, but van der Ree (1999) reported that these fences are effective for most stock, if tensioned. Barbed wire on the top two strands can entangle flying and gliding animals (Lindenmayer 2011). Alternatively, replacing the top one or two strands with high-tensioned plain wire will reduce most, but not avoid all, entanglements (van der Ree 1999; Booth 2007a).

On existing fences that are too old to withstand re-stringing, barbed wire can be covered or marked to increase its visibility to fauna. All-nylon highly visible sighter wire is nylon tape that is intended to increase the visibility of a barbed wire fence and has been shown to be effective, but is relatively costly (Booth 2007a).

Table 8.8.   Some alternatives and modifications to barbed wire fences

Type	Description
High-tensioned plain wire	Plain wire (i.e. unbarbed) tensioned for strength. All strands in the fence can be high-tensioned plain wire, or at least the top two strands.
Electric fences with no barbed wire	Strands in the fence are electrified. The remaining strands may be plain wire.
Emphasising the presence of a barbed wire fence	Marking barbed wire with a visual deterrent to increase its visibility and make it easier to avoid.
Covering barbed wire with split tubing	Physically covering the barbed wire so fauna are not caught on the barbs.

Box 8.6.   Virtual fence technology – cutting edge research reducing dependence on stock fences

A virtual fence is an innovative technology that, in the future, may replace the need to build conventional fences to contain livestock. Virtual fences do not involve any fixed wires; instead, livestock are fitted with a tracking collar that uses radio signals and global positioning satellites to generate electric and/or audio cues (Figure 8.26). These cues prompt the animal to stop when it approaches a boundary (Bishop-Hurley *et al.* 2007; Kubik 2007; Lee *et al.* 2009).

Figure 8.26.   Cattle fitted with a tracking collar that uses radio signals and global positioning satellite to generate electric and/or audio cues (Photo: CSIRO)

The CSIRO has begun trialling the technology in Australia and has reported promising results (Bishop-Hurley *et al.* 2007; Lee *et al.* 2009). Virtual fences could revolutionise future land management in Australia. Besides the obvious advantages of not needing to construct conventional fences, which can pose a hazard to wildlife, other foreseen benefits of the virtual fence include reduced farmer labour, as well as an improved ability to rotate stock to prevent over-grazing (CSIRO 2011). Similar to most technological advances, the current cost of installing a virtual fence limits its immediate widespread use (Kubik 2007).

Because most entanglements occur at night (Booth 2007a), reflective deterrents that operate in low light, such as reflective metal (Figure 8.27), are most likely better. Further discussion on the use of visual deterrents is provided in Chapter 7. Covering barbs with split tubing can be very effective because the barbs no longer present a risk.

Figure 8.27.   Aluminium cans incorporated into an old existing barbed wire fence, Dubbo New South Wales (Photo: J Gleeson)

Impacts on other aspects of the environment

Alternatives to barbed wire fences may still hinder the free movement of some animals. The risk can be greater for newly installed fences, particularly those that run across fauna movement paths (e.g. across water bodies or along ridges). Reducing the number of fences can help further reduce the impacts on fauna.

The addition of an electrified strand may improve retainment of stock, but electric fences can also pose a risk to animals (e.g. echidnas and flying foxes) so they need to be used with caution. For example, Stathan (2009) recognised that electric wires used on lower portions of a fence can cause deaths of echidnas as they put up their spines in response to the electric charge and become trapped.

Measures to minimise habitat degradation near a development site

Adverse impacts on wildlife are rarely limited to the direct disturbance footprint of a development. Habitat in close proximity to a development can be at increased risk of degradation. In the first instance, habitat changes (e.g. floristics and/or vegetation structure) occur around the perimeter of cleared habitat due to altered environmental conditions (e.g. increased sunlight or temperature changes). These altered environmental conditions occur at the edges of fragmented habitat and this is just one reason why avoiding habitat fragmentation is important.

The edge effects created from habitat clearance can be exacerbated to varying degrees by emissions from a development (e.g. sound, light and dust). Environmental weeds and exotic animals can further degrade the habitat as they make use of altered environmental conditions. Habitat degradation can also occur via changes in groundwater, surface water, noise or soil.

It is best to manage a development in such a way as to contain impacts as much as possible to the site (e.g. by adding controls on sound and light sources), rather than to rely on measures to rectify habitat degradation. For example, to reduce noise substantially, the noise source needs to be reduced or sited in a location that limits the spread of sound. Other measures commonly applied to reduce noise, such as conventional screens, will only slightly lower noise impacts on wildlife, if at all.

This chapter focuses on measures to minimise habitat degradation caused by environmental weeds, exotic vertebrate animals and light. Measures to tackle potential habitat degradation via groundwater, surface water, noise, soil or dust are outside the scope of this book, as discussed in Chapter 1.

Cumulative impacts on habitat in urban areas in particular can be considerable, rendering areas unsuitable for many species and reducing biodiversity. Urban encroachment results in increasingly blurred boundaries between developed areas and natural areas (Figure 9.1).

The placement and design of a development are important considerations in seeking to avoid impacts from development emissions that cannot be contained (e.g. using natural topography to shield emissions from a development). Setback distances are commonly applied to reduce edge effects on natural areas (Chapter 4). However, the extent of habitat degradation that may result from a proposed development is not easily predictable or quantifiable because a number of factors need to be taken into account. These factors include the impact pathway, the characteristics of the development, the topography, the characteristics of the surrounding habitat and the presence of susceptible species. There are also many gaps in the scientific

Figure 9.1.   Urban encroachment can act as a catalyst for the degradation of natural areas due to weed and exotic animal incursion and/or sound and light pollution (Photo: J Gleeson)

knowledge and understanding of how, and to what degree, particular species are tolerant of environmental weeds, exotic vertebrate animals and light.

Weed prevention and control

Overview	
What:	Preventing weeds from establishing within and near a development and controlling outbreaks.
Where:	Most development sites and associated activities can harbour or transport weeds.
Pros:	Prevents environmental weeds from establishing and/or spreading and degrading natural ecosystems.
Cons:	Some methods of weed control can harm other aspects of the environment or kill non-target species.
Cost:	Low–high (depending on the scale of, or potential for, weed infestation).
Companion measures:	Revegetation and the restoration of ecosystems (Chapter 10).
Alternative measures:	Avoid the development (Chapter 4).

Description and application

'Weed prevention' involves actions that prevent the incursion and establishment of weeds and 'weed control' involves eliminating or controlling outbreaks, if and when they occur. 'Environmental weeds' (also referred to here as weeds) are plants that invade and degrade natural

Table 9.1.   Examples of weed preventative measures

Type	Description
Avoid creating habitat for weeds	• Modify the development layout so that habitat clearance and fragmentation is minimised to essential areas only. • Ensure that the site surface water management is efficient to prevent changes in soil moisture that favour weeds. • Maintain the resilience of natural communities. • Landscape using flora species of local provenance. • Use physical barriers to prevent weed establishment in landscaped gardens, including natural (e.g. bark chips) and synthetic (e.g. woven cloth). • Encourage native vegetation cover to outcompete weeds.
Avoid spreading weeds	• Taking care when moving equipment and vehicles that could carry weed seeds or vegetative material. • Clean down equipment in designated areas. • Identify areas of environmental weeds and minimise access to these areas.

ecosystems, for example, by out-competing local native species. It is widely accepted that weeds are a considerable threat to nature conservation outcomes, as well as an economic problem worldwide, and should be prevented and controlled. In fact, landholders and occupiers in Australia have a legal obligation to control noxious weeds. If weeds are allowed to establish within a development site then they could spread into habitat in close proximity to the development and degrade it.

Weed prevention

It is important to understand how development can make an area vulnerable to the incursion and establishment of environmental weeds in order to prevent outbreaks from occurring. Vulnerability can be increased by:

- changing environmental conditions, such as land clearance that creates a bare substrate (weeds are able to quickly exploit bare areas) and poor site drainage, which can favour weedy species
- inadvertently importing seeds or vegetative material on vehicles or machinery required for the development
- planting environmental weeds in improved pastures or landscaped gardens.

It is generally more cost-effective to prevent environmental weed infestations from establishing in the first place than to control weeds once they become established (Hobbs and Humphries 1995). Weed prevention requires an understanding of which weeds are relevant to the development, their habitat requirements and their mode of dispersal. Weed preventative measures broadly fall into two groups: those that avoid creating habitat for weeds and those that prevent the spread of weeds, as described in Table 9.1.

Weed control

Weed outbreaks can still occur despite using weed preventive measures, particularly because some of the vectors of weed dispersal cannot be controlled (e.g. birds, wind and water). Early detection and control of weed outbreaks is a necessary component of an effective weed management program. Prerequisites for successful weed control include strategic planning, a sound knowledge of the biological and ecological aspects of the weed and the use of appropriate control methods. Some weed control methods are outlined in Table 9.2.

Table 9.2.   Weed control measures

Weed control measure	Description
Herbicide	Chemical substances designed to kill weeds, usually by interfering with growth processes.
Slashing or mowing	Slashing or mowing weeds before they flower to prevent them from setting seed.
Mechanical	Hand tools and/or machinery (e.g. tractors and earthmoving equipment) to physically remove weeds.
Grazing	Grazing by livestock (e.g. sheep, goats, cows and horses) to reduce weed growth and seed production and/or to prevent weed domination.
Fire	Controlled burning regimes to manage woody weeds over larger areas of land or targeted flame weeding for specific plants.
Biological control	Natural herbivores or pathogens of the weed introduced to control weed populations.
Allelochemicals	Growing native plants that exude or emit chemicals (allelochemicals) to inhibit the growth of neighbouring weeds.
Thermal weed control	Steam pressurised heated water applied directly onto weeds to break down cell structure. Hot water causes the loss of the waxy coating on the weed, a loss of moisture and subsequent dehydration.

The topic of weed control is large and a thorough treatment of all aspects is beyond the scope of this book. Information regarding the control methods recommended for specific weed species, including herbicide application rates, is often available on fact sheets or in reports produced by government departments (e.g. Ensbey 2009) or in scientific books (e.g. Parsons 1992; Ermert 2001). These guides can assist in planning effective weed management. People undertaking any weed control measures should be aware of their legal responsibilities. Anyone applying herbicides, for example, must follow the directions such as the rate, timing and application and safety on the label of the herbicide.

Effectiveness

Weed prevention or control can be considered effective if an area remains, or becomes, free of weeds or if weeds are contained at an acceptable level. The type of method best chosen for an application is situation-dependent. Because weed control measures are aimed at destroying plants, many of these measures can be effective, but can also present a risk to other environmental aspects (see section below).

Specific measures may be more effective on specific weeds, although it is likely that a combination of weed control measures (i.e. an integrated weed management approach) will be required to successfully prevent and control multiple weed species. Integrated weed management combines a number of appropriate weed control options (e.g. chemical, physical, fire and biological control methods) with an understanding of the ecology of the target species (Sindel 2000; Vitelli and Pitt 2006; Ensbey 2009). A coordinated program with other surrounding developments may be required to control weed sources and lessen the need for ongoing weed control.

Measures to prevent weeds from establishing form a strong basis for an effective weed management program. Replacing weeds with native species to encourage competition can, if done correctly, have valuable long-term benefits in restoring natural habitat resources for native wildlife. Using physical barriers to prevent weed establishment can be effective, but most perennial weeds can penetrate particulate mulches (e.g. sawdust) (Parsons 1992). Mulches also need to be replenished periodically.

When a weed outbreak is to be controlled, many people turn first to herbicides. Herbicides work effectively because the chemical kills the entire plant, including the roots below ground. This is an advantage over other control measures (e.g. slashing, mowing or grazing) that only affect plant parts visible above the ground surface where frequently repeated treatments maybe needed to be effective (Kristoffersen *et al.* 2008). Some herbicides, and their application techniques, can have adverse impacts on the environment (see below) and manual control measures (slashing or mowing) are common alternatives. Slashing or mowing reduces the above-ground component of a weed. These manual control techniques usually do not eradicate weeds, but they can assist native flora species to better compete (Pratley 2000).

In rural areas, livestock grazing is often used to suppress weed growth. Grazing livestock can vary in effectiveness depending on the palatability of the weed species (Ensbey 2009). It is likely that livestock grazing alone is insufficient and a combination of weed control measures will be required for successful weed control (e.g. herbicide application to remove unpalatable weeds).

Fire is used on occasion to control woody weeds. Of course, whether a fire will remove a weed depends on the intensity of the fire and the weed involved (e.g. a fire can sterilise weed seeds if they are exposed to high temperatures; Pratley 2000). A single fire generally will not control woody weeds on its own, but repeated fires or the integration of fire with other weed control methods can be effective (Vitelli and Pitt 2006). This is not always the case, however, and low–moderate fire may in fact promote the spread and persistence of some woody weeds (Gentle and Duggin 1997).

Biocontrol is used to target specific weeds. Some biocontrol agents have been shown to be very successful, such as the well-known use of the cactoblastis moth (*Cactoblastis cactorum*) to control prickly pear (*Opuntia* spp.) in Australia (Ermert 2001) (Figure 9.2). Established biocontrol agents are used widely and effectively in Australia for specific weed species. Unfortunately, established biocontrol agents are in the minority and biocontrol failures considerably outnumber successes (Julien 1992).

Figure 9.2. Cactoblastis moth (*Cactoblastis cactorum*) feeding on prickly pear (*Opuntia* spp.) (Photo: M Gleeson)

Another sophisticated form of weed control is using allelochemicals to inhibit the growth of specific weeds. Allelopathic (biochemical) interactions between plants are thought to influence flora species distributions within natural plant communities, and some weed species have been shown to use allelochemicals as a competitive strategy over native species (Rice 1977). For example, the weed lantana (*Lantana camara*) is known to suppress growth of other flora species (Gentle and Duggin 1997). Research is being carried out on these allelochemicals from weed species to aid the development of commercially available and effective herbicides.

Thermal weed control is a relatively new method and is still undergoing trials. Overseas, Kristoffersen *et al.* (2008) tested the efficacy of flame, steam, hot air and hot water weed control techniques and found that, although all of these methods reduced weed cover, hot water was marginally the most effective (although not significantly better than hot air or steam).

Impacts on other aspects of the environment

Weeds can have adverse impacts on the environment, but some weeds do provide habitat and food for native flora and fauna. Weed control measures applied to invasive plants with fleshy fruit, for example, can adversely affect native frugivores (Date *et al.* 1991; Gosper, 2004). This impact could be ameliorated by replacing the weeds with native plants that provide a comparable food source (Gosper and Vivian-Smith 2009).

Preventative measures can sometimes impact other aspects of the environment. Using physical barriers to smother weeds, for example, can suppress native flora species along with weeds. Also, if natural mulches are used, care should be taken not to introduce weed propagules. If they are used in natural areas (other than landscaped gardens), it should be noted that soil chemistry, soil conditions and soil micro-fauna may be affected by prolonged mulching (CRC for Australian Weed Management 2004).

The measures used to control environmental weeds must be target specific and not adversely impact any other native wildlife. Controlling weeds in natural areas can be challenging for this reason, and as outlined below, most methods have potential adverse impacts on other aspects of the environment.

Herbicides containing glyphosate can cause substantial mortality of amphibian larvae if mixed with a polyoxyethylene tallow amine (POEA) surfactant, which helps the glyphosate to penetrate the plant cuticles (Relyea 2005, 2006). Amphibians are particularly susceptible to the toxic effects because higher concentrations of these chemicals may accumulate in their preferred habitats (Mann *et al.* 2009). Weed populations have been known to develop resistance to such herbicides (Preston 2000).

Manual control measures can also have adverse impacts. Using tractors and earthmoving equipment in weed-infested areas can cause weeds to spread. Soil structure can be damaged and erosion may occur (Parsons 1992). Desirable native vegetation can also be considerably damaged or destroyed in the process.

In some situations, grazing livestock may actually help to spread weeds (e.g. burrs) and degrade native vegetation. Excessive grazing can render desirable species (e.g. native species) vulnerable to being out competed by weeds (Friend and Kemp 2000). Using grazing livestock to control weeds can damage native vegetation and soil structure, elevate nutrient levels attributable to their droppings, introduce weed species and encourage weed growth (CRC for Australian Weed Management 2004). However, some of these impacts can be avoided by adopting a low-intensity grazing program (McIntyre *et al.* 2002).

Particular fire frequency and intensity are needed to maintain vegetation communities and species in the landscape (Tolhurst *et al.* 1992; Driscoll *et al.* 2010). By maintaining natural fire regimes, the loss of a plant population may be avoided and, for this reason, caution is needed when attempting to use fire to control weeds due to the ecosystem requirements for particular fire frequencies and intensities.

Discouraging exotic vertebrates

Overview	
What:	Discouraging exotic animals from inhabiting a development by minimising attractants.
Where:	Most developments can attract exotic animals.
Pros:	Helps prevent exotic animals from competing with, or preying on, native species and otherwise degrading native ecosystems.
Cons:	Native species can also be discouraged, which may not be desirable in certain situations.
Cost:	Low
Companion measures:	Revegetation and the restoration of ecosystems (Chapter 10).
Alternative measures:	Avoid clearance of native vegetation and habitat fragmentation (Chapter 4).

Exotic vertebrate animals (often called 'introduced animals') can be discouraged from a development by minimising attractants, or restricting access to attractants. Developments can attract exotic vertebrate animals through the inadvertent creation of habitat resources. Food sources, such as refuse bins without lids, for example, can attract and help to sustain exotic vertebrate animals. We use the term 'discouraging exotic animals' because this chapter does not attempt to describe or evaluate options for the direct removal of exotic animals (e.g. trapping, poisoning and shooting).

There is little doubt that exotic invertebrates can also have a large impact on Australian wildlife, such as the impact of red imported fire ants (*Solenopsis invicta*) on ground fauna (Moloney and Vanderwoude 2002); however, the measures described in this chapter are not applicable to exotic invertebrates. Similarly, the investigation of hygiene and control measures to prevent the spread of diseases that affect wildlife, such as the amphibian chytrid fungus (*Batrachochytrium dendrobatidis*) – Box 10.3), is outside the scope of this book.

Often, exotic vertebrate animals exist in the general area regardless of the development activities (e.g. cane toad). However, developments can exacerbate existing problems because exotic animals can proliferate within a development site and spread into adjacent natural habitat. Generally speaking, exotic animals cause some of the greatest impacts on Australian wildlife. A number are naturalised – often referred to as 'feral animals', such as the rabbit and red fox – and are widespread in the Australian landscape. The impacts of exotic vertebrate animals depends on the particular species, but can include general habitat degradation (e.g. rabbit burrows can cause erosion), predation (e.g. foxes have decimated many species of small native marsupials on mainland Australia; Short and Smith 1994) and competition for resources (e.g. hollow logs are needed by echidnas and used by foxes).

Exotic animals can be discouraged by minimising attractants, restricting access to attractants and by encouraging local wildlife (Table 9.3). This requires an understanding of what exotic species are likely to be relevant to a development, their habitat requirements, their mode of dispersal and how the development may result in their incursion or proliferation.

Exotic animals (e.g. house mice) can be inadvertently transported on vehicles or machinery between development sites. Care should be taken to minimise this from occurring. For example, transported material should be inspected to confirm that there are no exotic vertebrate animals on board.

Table 9.3.   Examples of measures to discourage exotic animals

Measure	Description
Minimise attractants	• Modify a development layout to minimise habitat clearance and fragmentation. • Minimise the construction of additional tracks, which are often used by introduced predator species. • Rationalise development components that may attract exotic vertebrate animals.
Restrict access to attractants	• Maintain diligence with minimising access, such as closing lids on refuse bins after every use. • Use physical barriers to exclude exotic vertebrate animals.
Encourage local wildlife	• Encourage resilience of natural communities. • Landscape using flora species of local provenance to provide habitat for native animals.

Effectiveness

Measures to discourage exotic vertebrate animals are often overlooked and more emphasis is placed on the control of animals once they have established in an area. Discouraging these animals to prevent outbreaks from occurring in the first place is a more proactive approach for some of these species. Without suitable habitat, many exotic animals will be far less likely to occur within an area. Removing all habitat for exotic animals is difficult, but removing certain aspects can reduce the chances that they will proliferate. Examples of attractants that can be minimised to discourage exotic animals from occurring in, or near, a development are listed in Table 9.4.

The concept of removing an exotic animal's access to habitat has been recently proposed to prevent the cane toad (*Bufo marinus*) spreading into Western Australia. Cane toads require constant access to water and fencing artificial water sources is viewed as a key way to prevent further spread (Williams 2011). Western Australia's Stop the Toad Foundation regularly use fencing to cordon off water bodies and catch toads that collect on the fence. Regrettably, flood waters can hinder efforts, enabling the spread of cane toads to continue.

Table 9.4.   Examples of attractants that could be minimised to discourage particular exotic animals

Exotic animal	Example of attractants that may be minimised
Fox	• Minimise areas that may be used as a den (e.g. restrict entry to the space beneath raised buildings). • Avoid fragmentation of habitats that may enable foxes access to native habitats. • Control rabbits, exotic rats and mice, because these animals present a food source for foxes, and follow up with control measures targeted at foxes.
European rabbit	• Maintain resilience of natural communities to reduce the ability of rabbits to build warrens. • Encourage native predators (e.g. owls).
Cane toad	• Cover or fence artificial water sources.
Black rat	• Maintain diligence with minimising access to food sources, such as closing lids on refuse bins after each use. • Minimise areas that may be used for nesting (e.g. restrict entry to the space beneath raised buildings). • Encourage native predators (e.g. snakes).
Feral cat	• Control rabbits, exotic rats and mice, because these animals present a food source for feral cats, and follow up with control measures targeted at cats.

Measures to discourage exotic vertebrate animals are rarely 100% effective and control measures (e.g. trapping, poisoning and shooting) are inevitability needed if an outbreak is to be controlled. Controlling the prey (e.g. rabbits, exotic rats and mice) of exotic predators (e.g. fox and feral cat) can lead to them switching prey to native species. For this reason, control measures targeting these exotic predators are often required even though their current prey of exotic mammals has been taken away. The topic of controlling exotic animals is extensive and a thorough treatment is beyond the scope of this book. Further information on the ethical control of exotic vertebrate animals can be found in Sharp and Saunders (2004) and associated code of practices (Department of Sustainability, Environment, Water, Population and Communities 2011b).

A considerable limitation to discouraging exotic vertebrate animals by minimising attractants and access to attractants is that native plants and animals may use the same habitat resources. For example, food resources may not be able to be removed if the exotic animals forage on native plants (e.g. goats and pigs). For these types of species, control measures are imperative to their management. That said, maintaining resilient natural communities can, in some situations, discourage exotic animals.

Understanding how exotic animals invade an area is important when tailoring preventative measures. For example, degrading native ground cover (e.g. through grazing livestock) can enable rabbits to move into the area and establish warrens. In turn, an increase in rabbits can lead to increases in predatory red fox and feral cats because rabbits often account for the majority of their diet. This example also emphasises how exotic animals can aid each other in changing a landscape and how controlling exotic animals lower in the food chain may discourage others higher in the food chain.

Australia has a history of using fences to control exotic animals for agricultural reasons (e.g. the State Barrier Fence of Western Australia built between 1901 and 1907. Fences have also been used to separate exotic vertebrates from native fauna in sanctuaries (e.g. Moseby *et al.* 2009).

Impacts on other aspects of the environment

Minimising attractants at a development site (e.g. artificial food sources) can discourage exotic animals and have little to no affect on native species in some circumstances. However, if native species depend on an attractant, whether it is natural or human-made, the impact on these native species would need to be assessed against the benefits of discouraging exotic animals.

Reduction of artificial lighting

Overview	
What:	Reducing artificial lighting to prevent impacts on native fauna.
Where:	Developments that require artificial lighting in close proximity to natural areas or attract fauna with potentially adverse outcomes.
Pros:	Susceptible animals are not adversely affected.
Cons:	Generally few cons. A screen to shield lighting can be a barrier to movement for some species.
Cost:	Low–medium. Savings can be made by improving lighting efficiency.
Companion measures:	Revegetation and the restoration of ecosystems (Chapter 10).
Alternative measures:	Choose an alternative location for the development or redesign the development to minimise impacts using buildings or topography to shield light (Chapter 4).

Description and potential applications

Artificial lighting is used in most developments, often to allow the development to operate at night or for security. Examples of sources of light associated with developments include stationary sources such as street, path and building lights, as well as moving sources such as headlights from road vehicles and trains.

Natural light at night is important for many species because it influences foraging success, reproduction, predator avoidance, navigation and circadian rhythm (Salmon 2003; Longcore and Rich 2004; Rich and Longcore 2006; Stone *et al.* 2009; Rodriguez and Rodriguez 2009). So, it is logical to consider that artificial lighting can influence these aspects (refer to potential impacts from artificial lighting described in Chapter 2).

Light pollution is known to have an adverse impact on Australian marine turtles by interfering with nesting and recruitment (Environment Australia 2003; Limpus 2008) (Figure 9.3). Shannon (2007) demonstrated behavioural changes in captive sugar gliders (*Petaurus breviceps*) under high-light conditions, potentially reducing foraging time and ability to avoid detection by predators. It is surprising that there are few other studies that have examined impacts from artificial lighting on Australian fauna. Perhaps one reason is that the impacts from artificial lighting can be difficult to distinguish from other edge effects caused by development. For example, Pocock and Lawrence (2005) found that, on average, traffic noise from a road extended an average of 350 m into a forest and traffic light was visible 380 m into the forest).

It is worth noting that not all effects of artificial lighting are detrimental. For example, certain types of artificial light may improve foraging success in wading birds (Santos *et al.* 2010) and increase availability of prey and thermoregulatory opportunities for reptiles (Perry *et al.* 2008). There is a lack of research on how these changes affect animal populations, particularly in Australia. It is common to observe bats forage on insects, such as moths (Frank 2006) attracted to lights, but this can also be undesirable if it disrupts their natural behaviour.

Figure 9.3. Nesting adults green turtles avoid lit beaches and hatchlings can be disoriented by night lighting (Photo: D Gleeson)

For example, Stone *et al.* (2009) found a change in flight and foraging behaviour of the lesser horseshoe bat (*Rhinolophus hipposideros*) due to artificial lighting in the United Kingdom. It can also lead to increased risk of predation (Longcore and Rich 2006).

There are different types of artificial light sources. Sky glow is created by many lights, reducing the number of visible stars and can disrupt both local and distant ecosystems (Rich and Longcore 2006). For example, it can disorientate night-flying species moving between habitats (Longcore and Rich 2006). This would be a cumulative issue and generally not caused by a single development. Illumination of an area is a more direct impact on habitat and animals surrounding a development site. The intensity of light would influence the amount of light received at a certain distance from the source.

In certain ecologically sensitive areas, measures to reduce impacts from artificial lighting may be insufficient, and choosing an alternative location for the development can be more appropriate, such as avoiding nesting beaches used by the green turtle (*Chelonia mydas*). If it is not possible to move the entire development away from the ecologically sensitive area, it may be possible to redesign the development to minimise impacts by using building structures or topography to shield light.

Although many development sites are now individually powered by solar energy, lighting is mainly produced from other sources that emit high volumes of greenhouse gases. In these cases, there is often a desire to lower energy consumption. Although lowering energy consumption is very worthwhile, in this chapter we focus solely on reducing the impacts of artificial lighting on fauna surrounding a development site. These two objectives often coincide, but do not always. For example, dimming lights during dusk and dawn conserves energy, but this is not likely to reduce impacts on surrounding fauna because, during these times, the presence of ambient light means that artificial lighting is not an issue. In other situations, impacts on surrounding fauna might be able to be reduced by lowering the height of light fittings, but more lights may needed to illuminate the same area, increasing the overall energy consumption.

Ways to minimise adverse impacts from artificial lighting fall into three broad categories: those that minimise light; those that confine light; and those that substitute light sources. These measures are listed in Table 9.5 and further discussed further below.

Effectiveness

Just as there are few studies on artificial lighting impacts on Australian fauna, there are even fewer on the effectiveness of artificial lighting modifications. The gauge of their effectiveness is generally limited to information on how the modification changes the characteristics of the light (e.g. whether light is successfully minimised, confined or substituted). This is obviously an area that needs to be researched further so that it will be possible to better evaluate when artificial lighting modifications are required and how effective individual artificial lighting modifications are at reducing impacts. Any artificial light modifications used should not compromise the capacity of the lighting to achieve its purpose, or otherwise the purpose of the lighting needs to be re-evaluated.

The need for artificial lighting modifications depends on the occurrence of potentially susceptible species and characteristics of the artificial lighting proposed. Surveying nighttime environments using spotlighting, call playback, bat surveys and time lapse cameras can help in the identification and understanding of potentially susceptible fauna in an area. In ecologically sensitive areas, a combination of artificial lighting modifications may be required to reduce potential adverse impacts to fauna.

It is common sense that reducing the number of lights and their location can effectively minimise artificial light while still providing sufficient lighting for safety and operational

Table 9.5.　Examples of measures to reduce artificial lighting

Type	Measure	Description
Minimise	Minimise the number of lights	Increase spacing of fixtures to reduce the overall amount of artificial lighting.
	Turn off lights when not required	Control when the lights are on and off, assisted by the use of sensors, timers or motion detectors.
	Flashing lights	Lights that flash intermittently can be used where constant light is not required.
Confine	Light directing fittings	Use light fittings that direct and confine the spread of light (Figure 9.4).
	Lower light fittings	Reduce the height of the lighting to confine the spread of light.
	Window and door coverings	Blinds, curtains, window tinting and closed doors can minimise the dispersal of interior lighting to outside.
	Embedded lights	Use lights embedded in the ground, rather than on poles.
	Screens	Block light using screens of vegetation or screens made from timber (Figure 9.5), concrete blocks, toughened glass, acrylic or concrete or steel panels. Earth berms, created by hilling earth or by retaining earth, may also be used to screen light.
	Non-reflective surfaces	Design surfaces of structures and ground coverings to be non-reflective so light glow is not directed towards natural areas.
Substitute	Personal portable torches	Substitute light fixtures with personal portable torches.
	Low-pressure sodium bulbs	Substitute other lamps with low-pressure sodium bulbs to produce longer wavelength light (yellow–orange).
	Lower intensity bulbs	Substitute high-intensity (wattage) bulbs with lower intensity bulbs.
	Light-emitting diodes	Change the intensity/wavelength of the light using light-emitting diodes (LEDs).
	Light filters	Fit filters to light fittings to produce longer wavelength light (yellow–orange).

requirements. Similarly, turning off lights when not required is obviously effective. Seasonal light restrictions (e.g. turning lights off in the breeding season of a particular species) can be investigated.

Some work has shown that fixed light sources may attract more birds than flashing or strobe lights (see Jones and Francis 2003), but this is likely to be context-dependent.

Confining the spread of light (e.g. using light-directing fittings, lowering lighting and door coverings) are simple measures that can generally reduce wayward light without compromising the quality of light where it is needed. As previously mentioned, if the spread of light at each fitting is confined, then more fittings may be needed to cover the same area; however, the intensity of the light could be reduced at each fitting to minimise energy consumption. Embedded lights have been used in Florida, USA, to reduce night light compared with traditional street lights (Salmon 2006). It is well established that non-reflective surfaces (e.g. dark colours) reduce the glow and spread of light.

Figure 9.4.   Light fittings that direct and confine the spread of light at Osprey House Environmental Centre north of Brisbane, Queensland (Photo: J Gleeson)

Figure 9.5.   An example of a solid barrier made from timber erected along a roadside to confine the spread of headlights (Photo: D Gleeson)

A buffer (e.g. setback, barriers, topography or vegetation) may help shield artificial lighting from susceptible species. Opaque solid barriers and earth berms (or natural topography) can block light if they are designed to suit the situation (i.e. suitably tall and long to physically obstruct the source light). Vegetative screens are generally not as effective at reducing light as solid screens. Screens and earth berms are used along some roads or railways to screen vehicle headlights because the height of headlights is, in many cases, lower than the screen. Solid barriers at least 1.4 m high are used to reduce light from vehicular headlights. Taller screens may be required depending on topography (e.g. headlights can sweep over a greater area on down-slope sections of road). Earth berms can be used in conjunction with solid or vegetative screens, if necessary, to further reduce the permeability of light. Vegetative screens are intended to reduce light permeability to a natural habitat and comprise retained stands of tall, dense, evergreen vegetation or planted stands of suitable vegetation.

Even light sources that are minimised and confined may still have ecological impacts (Rich and Longcore 2006). Substitution may be an option to further reduce risks. Substituting fixed lights for personal portable torches can reduce artificial lighting substantially and still enable sufficient light where required. This may not be viable for a development with many people.

Different light bulbs have different light characteristics. Substituting light bulbs (e.g. lowering their intensity or changing the bulb colour) to effectively reduce impacts requires an understanding of how the target species will respond to the changed light characteristics. Significantly lowering the intensity of the light by substituting the bulbs may reduce impacts. For example, Jones and Francis (2003) demonstrated that lowering the intensity of a lighthouse beam could reduce adverse impacts on migratory birds.

Many species are least attracted to longer wavelength light in the yellow-orange to red end of the spectrum (630–700 nm). For example, yellow–orange light produced by low-pressure sodium bulbs attract fewer insects (e.g. moths) than white light produced by mercury vapour lamps (van Tets et al. 1969; Frank 2006). An additional benefit of low-pressure sodium lamps is that they are more energy efficient (120–175 lumens/watt). Although they are widely used in park landscaping and commercial buildings (e.g. service stations), they are generally considered unsuitable for lighting roads because people are not able to see yellow light as readily, and yellow light has low reflectance (Commonwealth of Australia 2007b).

Impacts on other aspects of the environment

Solid screens inhibit fauna movement in the same manner as fauna exclusion fences (Chapter 7). In some cases, such restriction of faunal movement may be of concern. For example, solid screens erected adjacent to roadways can cause fauna to become trapped on the road and put them at risk of vehicular strike. If solid screens have the potential to trap fauna, holes can be incorporated into the design of the screen. If solid screens are extensive and likely to divide a population or disconnect important habitat, the use of land bridges and/or fauna underpasses may need to be considered. Alternatively, earth berms can be used where fauna passage must be maintained, providing they are not too steep. Non-palatable plants can be considered for use in vegetative screens along roadways to discourage grazing fauna from coming too close to the road (discussed further in Chapter 7).

Low-pressure sodium bulbs need to be used with care near coastal nesting beaches used by marine turtles because hatchlings are attracted to the light they emit (Limpus 2008).

Measures to provide additional habitat for flora and fauna impacted by development

As discussed in Chapter 2, land clearance can result in the loss and fragmentation of habitat for flora and fauna. If species surrounding a development site are left with limited habitat resources, the size of local populations could fall, leading to greater risk of local extinction. The loss of habitat that occurs each time an individual development goes ahead can eventually lead to 'death by a thousand cuts', where a major change, such as the loss of a species, happens slowly in small increments that are not perceived as objectionable on their own. In the same respect, sound management of flora and fauna habitat at each individual development site will help maintain species across the landscape.

It is better to design a development to avoid and retain natural habitat resources than to attempt to replace them, because the alternatives can be inherently inferior. For example, it is always preferable to retain trees containing natural hollows because artificial tree hollows, such as nest or roost boxes (Figure 10.1), require continual maintenance. Additionally, many

Figure 10.1. A rainbow lorikeet in an artificial hollow at Dowse Lagoon, Brisbane (Photo: R Whitney)

specialised habitats cannot be re-created sufficiently well, or at all. Notwithstanding, providing substitute habitat components for fauna directly impacted by development can be particularly important. Resident fauna can become displaced and may not survive or reproduce because of a shortage of available habitat resources in the land surrounding the development. Habitat resources may simply be absent, or may be present but other animals are already using them.

Revegetation and restoration of ecosystems is discussed first in this chapter because this is a primary measure for putting flora and fauna habitat back into cleared and degraded landscapes. The other measures described in this chapter relate to specific habitat components (e.g. feeding, breeding, nesting and shelter resources) of various faunal groups. In this chapter, we also describe some innovative and value-adding measures that can assist in adding missing or sub-optimally prevalent habitat components for many types of fauna. These measures are particularly useful for promoting the recolonisation of areas that have been previously disturbed by developments.

Revegetation and restoration of ecosystems

Overview	
What:	Re-establishing native vegetation and restoring ecosystem function to provide wildlife habitat.
Where:	Within or near a development site.
Pros:	Wildlife is more likely to remain around a development site if suitable habitat is available. Assisted revegetation can also encourage wildlife to move back into an area sooner following temporary disturbance (e.g. after an area is mined).
Cons:	Revegetation and ecosystem restoration takes many years and some critical habitat components, such as tree hollows, may not develop for decades.
Cost:	Medium–high
Companion measures:	Plant salvage (Chapter 6); natural habitat linkages (Chapter 8); weed prevention and control and discouraging exotic vertebrates (Chapter 9); artificial tree hollow, artificial retreats for reptiles and salvaged habitat features (all this chapter); environmental offsets (Chapter 12).
Alternative measures:	Environmental offsets (Chapter 12).

Description and potential applications

Revegetation is discussed in this chapter as the deliberate planting, seeding or regeneration of plants for conservation purposes. Revegetation is often undertaken as a component in the complex process of restoration of an ecosystem. It is a complex process because it involves restoring all aspects of a natural ecosystem, such as species diversity, structure and ecological processes (Ruiz-Jaen and Aide 2005). Within the confines of this book, it is not possible to explore ecosystem restoration in depth, but it is covered briefly later in this chapter.

Land clearance and habitat degradation have produced highly modified landscapes in many parts of Australia. It is little wonder that the long-term viability of many native flora and fauna populations are at risk because of loss of habitat. As described previously, avoiding the need to clear native vegetation is a high priority (Chapter 4) and attempting to replace natural ecosystems with reconstructed ecosystems should only be attempted where impacts are unavoidable (Wilkins *et al.* 2003). However, because land clearance is an ongoing threat (Beeton *et al.* 2006), the re-establishment of native vegetation and restoration of ecological communities are major components of wildlife conservation in Australia (Attiwill and Wilson 2003; Carr *et al.* 2008; Paton and O'Connor 2010).

There are many different objectives and realised benefits of revegetation, such as habitat for flora and fauna, soil retention and water quality management (e.g. Abel *et al.* 1997). In this chapter, we consider undertaking revegetation for the provision of flora and fauna habitat. The re-establishment of native vegetation near a development site can provide food, shelter and breeding habitat resources for wildlife, reduce pressure on local wildlife populations and buffer existing wildlife habitat from the indirect impacts of development. Using revegetation to promote the safe movement of wildlife around a development site is discussed in Chapter 8 (e.g. wildlife corridors and stepping stones) and Chapter 12 describes revegetation and restoration of ecosystems in relation to environmental offsets.

The most appropriate revegetation is undertaken to recreate vegetation cover that existed in the area prior to clearance (Lindenmayer *et al.* 2003). Natural regeneration from existing seed sources (e.g. soil, canopy or water) followed by active management (e.g. supplementary planting from tubestock, as shown in Figure 10.2, or thinning for structural complexity) is the preferred method of re-establishing vegetation in natural areas (e.g. Williams 2000; McIvor and McIntyre 2002; Spooner *et al.* 2002). The level of active management required would depend on the resilience of the area, as discussed later. For example, rehabilitation of post-mine landforms may rely on tubestock or seed, particularly where there is deficient topsoil with a viable seed source.

Detailed methods on how to undertake revegetation are not described in this book. Instead, the reader can find this type of information in Greening Australia's *Native Vegetation Management Tool* (Carr *et al.* 2008), which is an online information source developed in collaboration with the CSIRO. An easy-to-understand introduction to ecosystem restoration is provided by

Figure 10.2.   Plant propagation for revegetation of post-mine landscapes (Photo: P Smith)

REGIONAL LEVEL

- Develop a long-term vision for the region
- Identify regional ecological priorities for revegetation
- Implement effective monitoring programs

LANDSCAPE LEVEL

- The amount of suitable habitat in the landscape
  - increase the total area of suitable habitat in the landscape
  - establish multiple populations
  - provide for species that use different habitats
- Enhance connectivity in the landscape
  - achieve connectivity by different configurations of habitat
  - give priority to streams and watercourses as natural corridors
  - recognise different kinds of movements through links
- Ensure representation of ecosystems
  - re-establish poorly represented habitats
  - restore remnants of depleted vegetation types

PATCH LEVEL

- Size
  - establish larger blocks for large populations
  - ensure habitats meet the area requirements of particular species
  - create larger patches for diverse animal communities
- Shape
  - increase width to reduce edge effects
  - design the shape and width of revegetation to meet species' requirements
- Location of blocks
  - position revegetation to increase opportunities for recolonisation
  - build on existing natural vegetation
  - locate new habitats away form known sources of disturbance
- Manage for diversity of vegetation

SITE LEVEL

- Use locally indigenous plant species
- Match plant species to the landform
- Establish natural layers in the vegetation
- Promote fine-scale patchiness of vegetation
- Provide ground-layer components as resources for wildlife and to assist restoration of ecosystem processes
- Management of vegetation and control of disturbances and degradation

Figure 10.3.   Principles for enhancing the value of revegetation for wildlife conservation (Source: Bennett *et al.* 2000, redrawn with permission from A. Bennett)

Buchanan (2009). Tongway and Ludwig (2010) and Peel (2010) also provide a good overview of landscape restoration in Australia.

Revegetation in Australia is occurring at multiple levels, from small-scale amenity plantings in urban landscapes to large-scale revegetation as part of regional corridors. An overview of the principles for enhancing the value of revegetation for wildlife conservation at different levels is provided in Figure 10.3. Development sites generally operate at a site and patch level, but these efforts are very important as they contribute towards the landscape and regional levels. These concepts are described further in Bennett *et al.* (2000). In addition to the site level components listed in Figure 10.3, missing habitat components (e.g. logs and hollows) can also be added at the site level to enhance revegetation efforts.

Revegetation is often undertaken with a focus on the whole ecological community, such as the restoration of grassy eucalypt woodland (Wilkins *et al.* 2003). However, revegetation efforts can also focus on a single species, such as the planting of *Allocasuarina* spp. as food trees specifically for the glossy-black cockatoo (*Calyptorhynchus lathami*) (Chapman 2007) or planting

Figure 10.4   Revegetation of koala food trees in Brisbane, Queensland. Note that the newly planted tube stocks are protected by plastic wind guards (Photo: J Gleeson)

of food trees for koalas (Figure 10.4). Providing targeted habitat resources for a particular species can be very worthwhile, though care should be taken not to overlook the needs of non-focal species (Lindenmayer and Fischer 2002; Ficetola *et al.* 2006). Nevertheless, the growing list of legislated threatened species in Australia tends to drive developers to revegetate to suit a particular species to avoid 'significant impacts' on that species. Such a strategy can also benefit a range of other species that use similar habitats.

Riparian areas along creeks and rivers are important habitat with high levels of structural and compositional diversity (McIvor and McIntyre 2002). Various authors have highlighted that revegetation efforts should prioritise drainage areas and watercourses (Fisher and Goldney 1997; Bennett *et al.* 2000). Revegetation in these areas benefit from improved soil development and water availability. Giving priority to revegetation of drainage areas and watercourses may also improve the functioning of the water catchment.

Revegetation and restoration of ecosystems is often undertaken on former agricultural land. Livestock grazing on natural vegetation continues to be the greatest land use in Australia (Australian Bureau of Statistics 2011b), dominating whole geographic regions. In the past, native vegetation was mostly considered to be of little value, besides stock shelter. Modern farms in Australia are now moving towards sustainable land practices that optimise both biodiversity and production outcomes, because of potential agricultural and biodiversity benefits. Native vegetation plays a crucial role in maintaining landscape function and productivity necessary for sustainable agriculture (Williams 2000; McIvor and McIntyre 2002). Nowadays, revegetation is part of general agricultural land management planning and practice. McIvor and McIntyre (2002) provides a list of principles for the sustainable use and management of eucalypt woodlands in agricultural landscapes. These principles include grazing livestock conservatively to maintain dominance of large and medium native tussock grasses, such as kangaroo grass (*Themeda australis*), and maintaining woodland patches greater than 5–10 ha to increase their viability (McIvor and McIntyre 2002). Generally, livestock should be excluded from watercourses to reduce soil erosion and maintain the quality of water (McIvor and McIntyre 2002).

In rural areas, fencing is used to control the entry of grazing livestock into areas undergoing revegetation. Fencing can improve flora species richness (Briggs *et al.* 2008) although some studies have found that grazing livestock do not necessarily need to be totally excluded from some vegetation types (e.g. grassy woodlands) to achieve high levels of species richness under temporary light grazing (Davidson *et al.* 2005). This is consistent with the well-known intermediate disturbance hypothesis that predicts that species richness will be higher at intermediate rates and intensities of disturbance (Connell 1978). Reduced grazing pressure can be achieved by rotational grazing or cell grazing (Lindenmayer 2011). Temporary light grazing is, however, more difficult to manage and regulate.

Effectiveness

There is little doubt that revegetation of cleared land can be an effective measure for providing habitat for flora and fauna (refer to Hobbs 1993; Kimber *et al.* 1999; Robinson 2006; Prober and Smith 2009; Munro *et al.* 2009). However, as discussed earlier, there are many objectives for revegetation, so it is important that the effectiveness of revegetation and ecosystem restoration is measured against clearly defined objectives (e.g. revegetation for a specific species or ecosystem restoration).

An assessment of ecosystem resilience should be conducted from the outset of revegetation and restoration activities because this will dictate the effectiveness of proposed restoration activities and level of active management required. This is basically a measure of the extent of degradation in the target area. How stable is the land? Are ecological process impaired and, if so, how are they impaired? Is there likely to be an existing native seed store in the soil based on the land use history? If the area is heavily degraded and it is unlikely there will be a viable seed store, management will need to be more intense to restore native flora diversity. In other situations, managing threats operating in an area (e.g. livestock grazing) can be enough to assist the recovery of an ecosystem. As previously mentioned, Tongway and Ludwig (2010) provide a good overview of landscape restoration in Australia.

Achieving a diverse understorey is recognised as one of the challenges for revegetation projects. For example, Munro *et al.* (2009) studied 26 revegetation plantings comprising approximately 20 tree, shrub and understorey species over a variety of temperate forest communities across West Gippsland, Victoria. The plantings ranged from 2 to 26 years old. The study found that revegetation plantings can achieve similar structural complexity as remnant vegetation within 30–40 years (not accounting for tree hollows and hollow logs). However, revegetation plantings of this age did not attain the ground layer floristic diversity of the remnant vegetation and appeared unlikely to do so without their deliberate introduction, because of the absence of recolonising smaller plants. Achieving a diverse understorey is important in many landscapes because habitat opportunities may be substantially increased by establishing a diverse understorey and varying plant density to cater to requirements of different bird and reptile species (Nichols and Nichols 2003). It is undoubtedly also important for other animals, including invertebrates. Methods of restoring understorey vegetation continue to be improved in Australia; for example, Cole and Lunt (2005) discuss restoring kangaroo grass (*Themeda triandra*) to grassland and woodland understoreys through different seeding techniques.

(a)

(b)

Figure 10.5. Revegetation of a mine on Stradbroke Island, Queensland. (a) A photo of a bare post-mine landform with a planted cover crop taken in 1999 and (b) the same post-mine landform in 2010 following revegetation and showing substantial growth and diversity in plants (Photos: P Smith)

Revegetation of cleared land takes many years and restoration takes even longer. The time lag to develop habitat values (Cunningham *et al.* 2007) needs to be considered in relation to the objectives of the revegetation, but it should not discourage the use of revegetation as a tool for wildlife conservation. Revegetation can provide a certain level of habitat in the short term, with habitat complexity increasing over longer periods (Box 10.1). In the short term, revegetation generally consists of grasses, herbs, shrubs and saplings. In the medium term, woodland and forest habitats develop mature trees with seed, nectar and/or lerp as a food source, a leaf litter layer and fallen logs. Tree hollows and hollow logs develop in the longer term. The 'before' and 'after' photos in Figure 10.5 show the result of progressive revegetation of the Stradbroke Island mine site over a 10-year period.

It is not surprising that different native species respond differently to revegetation, depending on the age of the regrowth and the habitat resources available (e.g. Majer *et al.* 2007). If the objective is to revegetate an area to provide the same habitat as remnant vegetation existing in the area, then missing habitat resources may need to be supplemented in the short to medium term. For example, habitat opportunities in revegetation areas can be enhanced by providing additional habitat resources, such as artificial tree-hollows, artificial retreats for reptiles and salvaged habitat features (which are all covered later in this chapter).

There are a myriad of factors that influence the success of revegetation and restoration efforts, including topography, hydrology, hydrogeology, soil properties (nutrient and carbon) and fire regimes. It is beyond the scope of this book to cover all of these aspects. Two main factors likely to influence the effectiveness of revegetation and restoration of ecosystems include the location chosen to revegetate (the placement) and ongoing management. These factors are discussed below.

Placement

In some situations, the placement of revegetation is pre-determined (e.g. revegetation of a disturbed area after a development). In other situations, there is a opportunity to choose the location in which revegetation efforts are focused. Location can have a large influence on the effectiveness of revegetation and restoration efforts. Plans to revegetate an area should be made considering current and future land uses. Will the particular restoration and revegetation activities still have the same anticipated value in the future?

Vegetation should be positioned to increase opportunities for recolonisation of native flora as well as to build on existing native vegetation (e.g. Paton and O'Connor 2010). This is also

Box 10.1.   Revegetation of land following mining

Mining is a temporary land use that involves extracting minerals out of the ground until the target resource has been exhausted. At that point, the land is rehabilitated for a new land use. The focus on revegetation of post-mine landforms is anticipated to intensify in the next 30 years in Australia, as many more post-mine landforms will need to be revegetated following the mining boom over the last decade.

The common objective for rehabilitation of mined areas is to re-establish native vegetation cover across stable post-mine landforms as a minimum, with more gallant objectives to restore functioning ecological communities. Due to the major environmental changes that are associated with open cut mining, it may not be possible to reinstate identical ecological communities to the pre-mining state; however, creating similar self-sustaining communities that fulfil a similar ecological role may be possible (e.g. Koch and Hobbs 2007; Tongway and Ludwig 2010).

Vegetation on post-mine landforms is traditionally perceived as simplistic monoculture vegetation with low habitat quality. Revegetating post-mine landforms with more complex and sustainable native ecosystems is a relatively modern concept developed over the past three decades and techniques are improving (see Nichols *et al.* 2005; Grant and Koch 2007). The success of mine revegetation is largely a function of how suited the physical environment is for the particular vegetation proposed to be grown (e.g. substrate and hydrology), as well as the amount of effort put into management. It is particularly important that revegetation proposals consider the inherent risks to success, so achievable goals are defined. Trained and competent personnel are also essential.

The effectiveness of revegetation on post-mine landforms can be improved by integrating the site into habitat in the wider landscape. Will revegetation integrate with other habitat in the landscape or become an isolated habitat patch? Surrounding habitat can act as a source of flora and fauna recolonisation and thereby improve the effectiveness of the rehabilitation process (Nichols and Nichols 2003).

Even with best practice rehabilitation design, it takes many years to restore diverse woodland and forest vegetation. People generally find it more challenging to grasp longer term goals because of inherent uncertainty as to what will happen in the future, particularly if it takes longer than a lifetime to achieve.

Although rehabilitated land with a sustainable native ecosystem with a full suite of species is more likely to occur in the distant future, more immediate successes are also possible. For example, in rehabilitating Jarrah forest on bauxite mine landforms in Western Australia, Nichols and Grant (2007) found that revegetation sites as young as 4–5 years can support similar bird numbers, densities and diversities as natural forest. In South Australia, Lee *et al.* (2010) recorded black cockatoos – Carnaby's black-cockatoo (*Calyptorhynchus latinostris*); Baudin's black-cockatoo (*Calyptorhynchus baudinii*) and forest red-tailed black-cockatoo (*Calyptorhynchus banksii naso*) – foraging on rehabilitating mine land within 8 years. In Queensland, research is showing that koalas are recolonising rehabilitating mined areas on North Stradbroke Island (Uhlmann 2010).

Table 10.1 Overcoming risks to revegetation success

Risk	Example of ways to overcome risks to revegetation success
Land instability, erosion and poor water transfer (high runoff)	Reshape the surface to create banks and troughs along the slope contour; incorporate salvaged vegetative material and logs to slow water transfer; plant dense seedings/tubestock; select plants that will hold soil; use mulch; construct sediment fencing
Poor water transfer (low infiltration)	Carry out deep-ripping (aerating) of the soil to reduce compaction; use mulch
Fire	Create a firebreak; control fire outbreaks
Drought	Irrigate; use mulch
Flooding	Re-position the revegetation activities to avoid high-flood risk areas; use earth bunding
Environmental weeds	Undertake weed prevention and control (Chapter 9)
Exotic animals	Discourage exotic animals (Chapter 9) and control them
Wind	Install wind guards; re-position the revegetation activities to avoid high-wind risk areas
Insect attack	Use insecticide; use biocontrol
Grazing	Install fencing and/or tree guards
Human intervention	Install fencing and/or signs
Lower flora diversity compared with pre-European vegetation community	Use topsoil with a viable native seed store; use ecological thinning to increase structural complexity; employ fire management; undertake soil amelioration; carry out additional seeding or planting
Poor seed germination	Re-evaluate species selection; re-evaluate the sowing technique
Poor regeneration of overstorey species due to high grass cover	Create a disturbance that reduces the grass cover before rainfall or before seed fall of an existing overstorey.

beneficial because more species often occur in larger patches of vegetation (e.g. Radford and Bennett 2010), among other reasons (e.g. increase core habitat) (e.g. Lindenmayer 2011).

Management

A program to monitor the success of revegetation or restoration activities should determine whether the approach needs to be adjusted in any way or whether alternative measures need to be incorporated. Appropriate ongoing management is required to achieve certain revegetation and restoration objectives (e.g. assessing the recovery of ecosystem resilience or restoration of landscape function).

Examples of measures used to overcome particular risks to revegetation success are provided in Table 10.1. One of the main ways to improve the effectiveness of revegetation and restoration projects is by giving adequate consideration to the site attributes when planning a revegetation program and providing an appropriate level of adaptive management and informative monitoring over a relatively long time period.

Impacts on other aspects of the environment

Plants are important components of hydrological cycles. Plant roots take up water in the soil and, via transpiration, water is expelled from the plant as water vapour. Thus, revegetation can influence hydrology and hydrogeology (e.g. revegetation can be used to lower rising ground-water to avoid dryland salinity issues) (Eamus *et al.* 2006)

Artificial frog ponds

Overview	
What:	Creating artificial frog ponds to provide habitat for frogs.
Where:	Useful where a development (e.g. residential) could destroy or significantly disturb a water body used by frogs.
Pros:	Provides breeding, refuge and foraging habitat for frogs and other animals.
Cons:	Usually requires ground disturbance. High maintenance may be required.
Cost:	Medium–high
Companion measures:	Fauna exclusion fences (Chapter 7); natural habitat linkages and fauna underpasses (Chapter 8); weed prevention and control and discouraging exotic animals (Chapter 9); Revegetation and ecosystem restoration (this chapter).
Alternative measures:	Avoid disturbing ponds or other low-lying areas used by frogs (Chapter 4) and wildlife-friendly dams (this chapter).

Description and potential applications

Artificial frog ponds are bodies of water specifically created to provide breeding habitat for particular frog species. Artificial frog ponds are increasingly being incorporated into coastal developments in eastern Australia.

The first development in the southern hemisphere to incorporate artificial pond habitat specifically designed for frogs was the highly publicised Sydney Olympic Park development in New South Wales. Here, artificial frog ponds were created for the conservation of a population of the threatened green and golden bell frog (*Litoria aurea*) (Pyke and White 1996) (Figure 10.6) (Box 10.2). Examples of artificial frog ponds created since include those specifically created for the threatened wallum sedge frog (*Litoria olongburensis*) along the Tugun Bypass in South-East Queensland and the southern bell frog (*Litoria raniformis*) on the

Figure 10.6. Two artificial frog ponds created for the green and golden bell frog at Sydney Olympic Park, New South Wales. Note the gradual sloping banks, emergent vegetation and the rocks that have been added for basking and refuge habitat (Photo: J Gleeson)

Figure 10.7.   (a) Created artificial frog pond, Pakenham Bypass, Victoria (Photo: A Organ) and (b) artificial frog pond, Tugun Bypass, Queensland (Photo: J Gleeson)

urban-fringe of south-east Melbourne, Victoria (Figure 10.7). Both the green and golden bell frogs and wallum sedge frogs are known to readily recolonise disturbed areas, suggesting that the creation of artificial habitat is possible for these species. Unfortunately, the prognosis may not be so good for frog species that rely on attributes of natural or undisturbed habitats (Pyke and White 1996).

Habitat loss is one of the main threats to frogs in Australia (Hazell 2003). The removal of natural ponds used by frogs can contribute to a gradual breakdown in the connectivity of frog populations. Artificial frog ponds are worthwhile considering to automatically provide habitat and to maintain connectivity of frog populations. It should be noted, however, that if frog habitat is proposed to be disturbed, the most appropriate action may not be to replace the habitat with a frog pond. The availability of suitable breeding habitat may not be the limiting factor to the persistence of the frog population in question. Instead, management and enhancement of the existing habitat may be more beneficial. Key threats to frogs in the area, such as predatory fish or the introduced cane toad (*Bufo marinus*), should be evaluated to determine if managing them will improve the viability of the local frog population.

Small-scale frog ponds have been constructed in backyards across Australia. They are made from all sorts of materials, including plastic pond liners and prefabricated ponds. These ponds can be successful for a range of frog species, such as the striped marsh frog (*Limnodynastes peronii*) and green tree frog (*Litoria caerulea*). Similar ponds have even been used experimentally at Sydney Olympic Park (Box 10.2). However, in this chapter we focus on artificial frog ponds constructed using natural materials (i.e. earth movement and provision of a clay base) because this design will be more sustainable over the long term, reducing costs and need for intense management. In some situations, raised ponds may be required for enhanced protection from predators (e.g. cane toads), but there are also other options available as described later in Table 10.2.

Artificial frog ponds can be different shapes and sizes, ephemeral or perennial. Frog ponds constructed along the Tugun Bypass for the wallum sedge frog are approximately 15–20 m long and approximately 5–10 m wide. Hamer and Mahony (2010) suggested that water bodies greater than 0.5 ha in surface area may be preferable for the green and golden bell frog.

The shallow hole for an artificial frog pond is ideally excavated during a dry period to enable machinery to access the site. The holes can often be left to naturally fill with rainwater. Swales and embankments can be created to help direct surface water flow to the ponds. After the dam fills, the sides of the pond are revegetated with aquatic and semi-aquatic plant species native to the area and known to be used as habitat by the target species. Adding refuge habitat

Box 10.2.   Creation of artificial habitat for the green and golden bell frog at Sydney Olympic Park, Sydney

Sydney Olympic Park was developed at Homebush in the late 1990s as a venue for Sydney to host the 2000 Olympic Games. A population of the threatened green and golden bell frog (*Litoria aurea*) existed within the development site, so a program to conserve this species was necessary. The concept was to retain and enhance core breeding habitat and also to create new areas of breeding habitat (Goldingay 2008).

Purpose-built frog ponds were constructed on remediated lands (Figure 10.6). A total of 70 frog ponds have been installed at Sydney Olympic Park, made from a variety of materials: some clay-lined, some rubber-lined and some made from fibreglass. The ponds were vegetated to a maximum of 80% vegetation cover with plants such as rushes and sedges (New South Wales Department of Environment and Climate Change 2008a).

The green and golden bell frog is also known to use a range of artificial water bodies, including dams in an industrial site at Kooragang Island, Newcastle in New South Wales. The general habitat preferences of this species contributed to the general success of the frog ponds at Sydney Olympic Park. However, the artificial frog ponds at Sydney Olympic Park are not self-sustaining. They require intense maintenance of the adjacent vegetation and control of mosquito fish (*Gambusia holbrooki*) (Darcovich and O'Meara 2008), as well as control of amphibian chytrid fungus (*Batrachochytrium dendrobatidis*) (Penman *et al.* 2008).

(e.g. rock piles) is important for shelter. A sustainable water source into the pond is often required; otherwise the ponds may provide only ephemeral habitat.

Sustaining a population of frogs at an artificial frog pond also requires provision of surrounding foraging, basking, refuge and movement habitat used by frogs. Natural habitat linkages are important for connecting habitats (Muir 2008; Hamer and Mahony 2010) (Chapter 8). Salvaged habitat features (discussed later in this chapter) may be added for basking and refuge habitat.

Effectiveness

The artificial frog ponds constructed at Sydney Olympic Park over a decade ago have proved successful by resulting in the establishment of two new sub-populations of the green and golden bell frog (Darcovich and O'Meara 2008) (Box 10.2). The long-term success of the artificial frog ponds built for the wallum sedge frog and the southern bell frog are so far unknown, because the ponds are relatively new (see Box 8.3).

Factors likely to influence the effectiveness of an artificial frog pond include its design, placement and maintenance. These factors are discussed below.

Design

There is no single frog pond design that will be optimal for all frog species. Different species generally have their own particular requirements for choosing suitable ponds. The key to creating artificial habitat for frogs is to understand the habitat needs of the target species, and to best mimic the ponds used by them (e.g. the wallum sedge frog breeds in wetlands with acidic soil). Where there is currently a lack of scientific studies on the successful installation of artificial frog ponds for a particular species, the measure should be considered experimental and monitored and evaluated for success or otherwise.

Artificial frog ponds should ideally be constructed with gradual sloping banks, to provide frogs with easy access to the pond and tadpoles with warmer shallower water near the edges in winter and cooler, deeper water in summer (Anstis 2002). Wetland plants are important components of artificial ponds, not only as habitat, but also to reduce nutrient levels, filter sediment from the water, oxygenate the water, regulate light availability for algal photosynthesis, decrease erosion and provide a basis for a healthy biological food chain and shelter from predators (New South Wales Department of Land and Water Conservation 1998; Anstis 2002).

Advice on suitable plants can be found in Romanowski (2009). Care should be taken when planting trees near artificial frog ponds because some leaves can be toxic to frogs. In other instances, shading can hinder growth of wetland plants and excess leaf material in the pond can lead to algal blooms. Some frogs have also shown preference for ponds that are exposed to direct sunlight for at least 6 hours a day during spring/summer (White and Pyke 2008).

Placement

When planning the placement of an artificial frog pond, there are a number of considerations. The primary consideration is whether the proposed location for the artificial frog habitat is within the range of the target species and whether the target frogs are likely to move into the habitat unaided. For example, Hamer *et al.* (2002) found that green and golden bell frogs were more likely to occupy ponds within 50 m of ponds already occupied by other individuals.

Some have suggested that artificial lighting (e.g. solar lights) may be used to encourage frogs to move into the newly created habitat. The idea is that the lights attract moths and other insects, on which frogs feed. However, this technique should be evaluated experimentally: overseas studies have shown that artificial lights can be detrimental to frogs being able to detect and catch prey (Buchanan 1993).

Self-sustaining artificial frog ponds depend on water from rainfall and local runoff. In a sense, the construction of a frog pond requires similar considerations as a farm dam in that it is desirable for the frog pond to have low siltation with minimal site runoff, flood water management and a base that is slow release and maintains perennial water. Lewis (2002) provides a good overview of farm dam planning and construction. Sometimes, site restrictions can mean that the natural water sources will not suffice. The hydrology of some of the artificial frog ponds built at Sydney Olympic Park is controlled by a water recirculation system (Darcovich and O'Meara 2008). In addition to maintaining water supply, the water recirculation system is used to control water levels and limit infestation by the eastern mosquitofish (*Gambusia holbrooki*) (O'Meara and Darcovich 2008).

Chemical, biological and physical processes are important considerations for an effective self-sustaining frog pond. For example, the water quality and chemical processes in a frog pond are influenced by the soil profile at the location and runoff from the surrounding catchment.

Maintenance

Management of vegetation is important for maintaining an effective artificial frog pond, because ponds overrun by vegetation (e.g. *Typha*) can be less attractive habitat for some species. The management required can be quite intense in artificial systems (e.g. Sydney Olympic Park), but is likely to be less so in more natural environments. It is important to choose water plants that are less likely to become problematic (e.g. overgrow through the pond). Management of environmental weeds can also be particularly important.

Table 10.2.   Overcoming risks to successful artificial frog ponds

Risks	Example of ways to overcome risks
Cane toads (where applicable)	Cull cane toads (trap and collect).
	Build an amphibian fence around the pond to exclude toads.
	Raise the pond (greater than 35 cm) from the surrounding ground level and remove climbing substrates to exclude toads.
Predatory fish	Isolate the pond from sources of predatory fish, such as eastern mosquitofish (*Gambusia holbrooki*); also keep in mind the potential for fish to be transported into the area during flooding.
	Manipulate the water levels.
Predatory birds	Use coverings such as overhead wires and netting.
Rats	Use control methods such as baiting and trapping.
Weeds	Remove weeds via mechanical means.
Blue-green algae	Plant aquatic and semi aquatic vegetation to reduce light, increase nutrient uptake and reduce particulate suspension.
	Recirculate the water and provide turbulence in the water.
	Control water quality.
Amphibian chytrid fungus	Follow hygiene protocols, such as those used by the New South Wales Department of Environment and Climate Change (2008b).

An artificial frog pond can be at risk of failing due to threatening processes operating in an area. Possible solutions to overcoming risks to the establishment of successful artificial frog ponds are provided in Table 10.2. Of course, not all of the potential risks will be applicable to all artificial frog ponds.

Impacts on other aspects of the environment

Because the creation of an artificial frog pond requires excavation, it is a development in its own right and the impact on the environment would need to be assessed. Consideration would also need to be given to the practicality of constructing a pond without impacting the existing habitat and movement corridors.

The artificial frog pond should be suitably located. For example, frog ponds proposed along a road may increase the susceptibility of the target species to road mortality, so additional measures may need to be considered, such as. exclusion fences (Chapter 7) and fauna underpasses (Chapter 8).

A high degree of effort is required to avoid transporting amphibian chytrid fungus into artificial frog ponds, because the fungus may then infect natural frog populations (Box 10.3). For example, Stockwell *et al.* (2008) attempted to re-introduce the green and golden bell frog into artificial frog ponds in the Hunter Wetlands, New South Wales and, although 850 tadpoles were released into the ponds, within 13 months the bell frogs appeared to have died out, presumably due to amphibian chytrid fungus.

Artificial frog ponds can provide a breeding ground for mosquitoes, if not controlled. Possible control methods include: introducing native fish that will prey on mosquitoes but not prey on frogs, such as pacific blue-eye (*Pseudomugil signifer*) and Australian smelt (*Retropinna semoni*); recirculating the water and fluctuating water levels during the mosquito breeding cycle; attracting invertebrates to eat the mosquito wrigglers by providing rocks and aquatic vegetation; installing sprinkler systems that inhibit mosquito colonisation and development; and target-specific chemical control.

Box 10.3. The importance of awareness of amphibian chytrid fungus

Amphibian chytrid fungus (*Batrachochytrium dendrobatidis*) is a pathogenic fungus that causes the infection known as chytridiomycosis. It was first introduced to South-East Queensland in the 1970s and subsequently spread along the east coast of Australia and south-west Western Australia (Department of the Environment and Heritage 2006). The fungus can cause an epidemic wave of high mortality, essentially devastating frog populations. A high degree of rigour is required to avoid transporting amphibian chytrid fungus into artificial frog ponds (Department of the Environment and Heritage 2006), because it can be spread by people moving through infected areas.

Penman *et al.* (2008) discusses how an outbreak of amphibian chytrid fungus was identified and managed at the Sydney Olympic Park. Considering the potentially devastating effects of the fungus on a population, Penman *et al.* (2008) suggests that maximising the variety and number of ponds available to the species, as well as ensuring connectivity between the ponds enabling healthy frogs to recolonise the habitat after the infection, may have helped the overall population withstand the outbreak of chytrid fungus. However, connectivity of habitat can also aid in the spread of amphibian chytrid fungus and a trade-off of the risks would need to be considered.

White (2006) investigated whether low salt concentrations in the pond water could kill amphibian chytrid fungus and first indications were positive. This study opens up the possibility that artificial frog ponds may be able to be designed in such a way as to be protected against amphibian chytrid fungus. There is no doubt, however, that further studies on possible techniques to fend off outbreaks of amphibian chytrid fungus are warranted.

Wildlife-friendly dams

Overview	
What:	Enhancement of an existing farm dam to provide habitat for wildlife.
Where:	Anywhere there is an existing farm dam (e.g. agricultural areas or rural residential areas).
Pros:	Additional habitat resource can encourage wildlife to stay in an area.
Cons:	Increases in undesirable species (e.g. common or introduced species) are possible.
Cost:	Low–medium
Companion measures:	Fauna exclusion fences (Chapter 7); alternatives to barbed wire (Chapter 8); weed prevention and control and discouraging exotic animals (Chapter 9); revegetation and ecosystem restoration and salvaged habitat features (this chapter).
Alternative measures:	Artificial frog ponds (this chapter).

Description and potential applications

Wildlife-friendly dams are water bodies that are modified to enhance their habitat value for wildlife. These enhanced water bodies can be viewed as a type of artificial wetland. With more than half a million farm or residential dams in Australia (Brainwood and Burgin 2009), they have the potential to play a vital role in wildlife conservation (Lindenmayer *et al.* 2003; Hazell *et al.* 2004). A wildlife-friendly dam could be used by a range of native species, providing habitat resources for birds, reptiles, frogs (Hazell *et al.* 2001), a variety of macroinvertebrates (Brainwood and Burgin 2009), and a water source for bats (Churchill 2008).

Water is a key resource for many animals and its presence or absence can dictate the types and abundance of fauna present in an area. This is particularly true for arid and semi arid areas of Australia, where artificial water sources are created primarily for use by stock (Landsberg *et al.* 1997; Fensham and Fairfax 2007; Wallach and O'Neill 2009).

Farm or residential dams come in all different shapes and sizes (e.g. ephemeral and small dams). Some dams provide limited opportunity for enhancement due to the nature of their construction (e.g. steep banked dams). If a new dam is to be installed, the characteristics that will generally make the dam more desirable to wildlife include the following (Hill and Edquist 1982; Winning and Baharrell 1998; Brouwer and Young 1998; Lewis 2002):

- A large surface area, and in particular, irregular edges that are vegetated, makes the dam more desirable to waterfowl.
- Dam margins that are shallow (25–400 mm), ephemeral and fluctuate with rainfall are desirable for waterfowl (e.g. ducks, geese and swans) and other waterbirds (e.g. herons, ibis and grebes), as well as some frog species.
- Dams with gradual sloping banks allow easy access for fauna, as well as substrate for plant establishment.
- Retaining topsoil from the dam excavation and respreading it over the clay base and sides of the dam can promote fertility and make it easier to establish vegetation around the dam edges.
- Locating the dam near natural bushland will provide resident fauna with a water source.

An existing farm or residential dam can be enhanced to provide improved habitat for flora and fauna via a number of measures described in Table 10.3 and further below.

Due to the disturbance involved in modifying an existing dam, consideration should be given to native species possibly already supported by the dam. If a dam already provides suitable habitat for native species, it may be more appropriate not to attempt any enhancements. For example, if the threatened green and golden bell frog (*Litoria aurea*) uses the dam, attempting to construct mud flats for birds could devastate the frogs' habitat.

In addition to the measures in Table 10.3, providing artificial hollows can help attract waterfowl that nest in tree hollows, such as the Australian wood duck (*Chenonetta jubata*) and Australian shoveler (*Anas rhynchotis*). This is discussed further in the next section.

Wildlife islands

Islands made from land in the middle of a dam ('land islands') provide birds with a haven from predators and habitat for roosting and nesting (Figure 10.8). The shallow water around the island can provide habitat for invertebrates and frogs. Land islands are created by pushing soil into the dam and then creating a channel. Hill and Edquist (1982) recommend the channel around the island should be at least 1 m deep and 3 m wide to keep most predators off the island.

An alternative to creating a land island is to use a floating island. Floating islands require little or no disturbance to an existing dam and are suitable for small dams. The island, generally 1–2 m², is constructed from buoyant materials (e.g. PVC water piping) and is anchored to

Table 10.3. Examples of dam enhancements

Type	Description
Wildlife islands	Small islands of land, or floating islands, in a dam can provide a haven for animals, particularly birds.
Mud flats	Low areas beside a dam that are periodically inundated with shallow water (approximately 30 cm deep) may provide habitat for wading birds.
Partially submerged upright logs	Logs that are fixed in an upright position in or around a dam may be used as a roosting perch for birds.
Retreats	Refuge habitat (e.g. logs and rocks) may be used by invertebrates, frogs and reptiles.
Revegetation	Plants in and around dams provide habitat for a range of invertebrates, frogs, reptiles, fish and birds, as well as habitat for other plant species (by altering the microhabitat).
Livestock management	Restriction or exclusion of livestock access to the dam stabilises the soil and protects vegetation.
Addition of fish	Stocking the dam with commercially available native fish will benefit animals that feed on them.

the base of the dam to prevent movement from wind and wave action. Floating islands can be topped with soil and vegetation to simulate a land island, but the practicalities of sustaining this in the long term can be challenging because of erosion.

Figure 10.8.   A wildlife island and mud flat at the Australian Botanic Garden, Mount Annan, New South Wales (Photo: J Gleeson). Note that the mud flat is in the foreground and a deep channel separates the island from the surrounding land

Mud flats

Mud flats can be created alongside farm or residential dams by excavating or infilling shallow areas (approximately 30 cm deep) (Winning and Beharrell 1998). The mud flats are mainly used by waterbirds such as bitterns and herons. Excavating or infilling should be done using suitable soil; otherwise water quality within the dam may be adversely affected.

Partially submerged logs

Tree logs that are fixed in an upright position and partially submerged within or beside a dam, can be used as a perch for roosting birds. Trunks with a number of fixed perches or branches at different heights will provide better roosting opportunities. Trees salvaged from development clearance areas can be used (discussed later in this chapter). Those placed within the water will be less susceptible to termite damage than logs placed around the outside of the dam.

Retreats

Piles of logs and rocks beside or in a dam can provide refuge habitat for invertebrates, frogs and reptiles. Reptiles will use logs and rocks beside a dam for basking. Different microhabitats can be created by adding a variety of substrates to the shoreline around the dam (sand, pebble, gravel or clay) (Winning and Beharrell 1998). Natural material for retreats can be salvaged from areas to be cleared for development or made from artificial materials (which are both discussed later in this chapter).

Revegetation

Plants are important components in the functioning of wetlands, as described in the previous chapter. Plants in and around farm or residential dams provide breeding, forage and refuge habitat for a range of invertebrates, frogs, reptiles, fish and birds, as well as habitat for other plant species. This includes sedges, rushes, reeds, emergent and sub-emergent plants and, further back from the edge of the dam, herbs, shrubs and trees. A greater variety of plants in and around the dam will generally equate to greater complexity of habitats. Advice on suitable aquatic and semi aquatic plants can be found in Romanowski (2009).

A lack of surrounding vegetation could be the greatest limiting factor to the habitat value of a farm or residential dam. The vegetation upslope helps trap sediment and nutrients before they are washed into the dam (Brouwer and Young 1998). Additionally, connecting a dam to existing habitat areas by revegetating a linkage (see Chapter 8) is likely to increase fauna usage of the dam.

Care should be taken not to plant trees on the dam embankment because this can lead to dam instability once the trees grow (Hill and Edquist 1982).

Livestock management

Many farm dams that are not used for irrigation are installed to provide water to livestock. Livestock access to a farm dam can significantly degrade the habitat value of a dam by increasing sedimentation, pollution and by damaging surrounding vegetation cover. Fencing can be used to manage livestock grazing near the dam, while still providing water for domestic use. Excluding stock from the dam entirely with fencing, and supplying stock with pumped (often powered with solar power) or siphoned water, is one way to avoid this impact (Voigt 2006). Fencing can be beneficial, but may be problematic if the access is over-trafficked and erodes.

If a fence is to be installed near a farm dam, it is important that it does not contain barbed wire because wildlife may become entangled in it (Chapter 8). If fencing is not appropriate, other options include reducing stocking rates or adopting a paddock resting regime to assist in maintaining vegetation cover.

Addition of fish

Adding native fish to a farm or residential dam can provide a food source for higher trophic levels of the food chain. It may be better to use native fish that will not prey on frogs, such as pacific blue-eye (*Pseudomugil signifer*) and Australian smelt (*Retropinna semoni*) (Grant 2003). Removing predatory fish from a dam, such as eastern mosquito fish (*Gambusia holbrooki*) may improve the habitat for frogs.

Effectiveness

An effective wildlife-friendly dam will provide habitat for wildlife. This can often be achieved without comprising the function of the dam for agricultural practices.

The availability of wildlife habitat in the catchment will improve the use of the dam by fauna. Vegetation in the catchment will also help reduce erosion and potential siltation of the dam. For example, in an extensive survey of 70 dams in south-eastern Australia, Hazell *et al.* (2001) found that farm dams with emergent vegetation cover around the margin, less surrounding bare ground, and a greater native tree cover within 1 km are most likely to have the highest conservation value for local frogs.

The potential risks to establishing successful artificial frog ponds and ways to overcome them are provided in Table 10.2. These are also relevant to a dam that has been enhanced for wildlife. Maintaining a wildlife-friendly dam will require regular inspections of: the dam stability; the health and vigour of plants; the need for fence maintenance; the state of wildlife islands; and the presence of weeds and exotic animals.

Impacts on other aspects of the environment

The increase in farm dams since European settlement is thought to have caused increases in native species such as the galah (*Eolophus roseicapilla*), sulphur-crested cockatoo (*Cacatua galerita*) (Lindenmayer *et al.* 2003) and kangaroos (*Macropus* spp.) (Fensham and Fairfax 2007) and increases in exotic animals such as the cane toad (*Bufo marinus*). Given that water in the landscape has such a large influence over fauna distribution, the effects of providing a new water source, or decommissioning an existing artificial source, can be complex. Hence, controversy remains over whether artificial water sources really do benefit biodiversity in parts of Australia (Fensham and Fairfax 2007; Wallach and O'Neill 2009).

Artificial tree hollows

Overview	
What:	Artificial tree hollows are box structures intended to provide substitute roosting and/or nesting habitat for fauna that use natural tree hollows.
Where:	Useful where hollow-bearing trees are to be, or have been, removed from a development site.
Pros:	Hollow-dwelling species continue to live and reproduce in an area.
Cons:	Potential to increase populations of common or introduced species. Not all hollow-dwelling fauna are known to use artificial tree hollows.
Cost:	Low–medium.
Companion measures:	Discouraging exotic animals (Chapter 9); revegetation and the restoration of ecosystems and salvaged habitat features (this chapter).
Alternative measures:	Avoid disturbance to trees containing hollows (Chapter 4).

Description and application

Artificial tree hollows (e.g. nest or roost boxes) can be provided as a habitat resource for hollow-dwelling fauna if hollow-bearing trees are to be, or have been, removed from a development site. By essentially substituting an otherwise lost habitat resource, land managers can encourage hollow-dwelling species to continue to live and reproduce in otherwise suitable habitat in the local area. Without this management action, the removal of hollow-bearing trees can result in a loss of diversity of hollow-dwelling species, with some species at risk of becoming locally extinct.

Removal of hollow-bearing trees should be avoided wherever possible, given the difficulties and costs associated with providing long-term, effective alternatives. Lengthy periods of time are generally required for hollows suitable for vertebrate fauna to develop. For example, eucalypts generally don't develop hollows suitable for vertebrate fauna until they are between 120 and 180 years old, and larger hollows are rare in eucalypts less than 220 years old (Gibbons and Lindenmayer 2002).

Areas of regenerating woodland and forest are usually deficient in hollows. Artificial tree hollows can be used as a temporary habitat resource while such an area is being restored to speed up the process of recolonisation by hollow-users. Although it takes many years, the saplings in the rehabilitated areas should eventually mature and develop natural tree hollows, so artificial hollows will no longer be required. The biodiversity of farm forestry plantations can also be enhanced by supplementing habitat with artificial tree hollows (Smith and Agnew 2002).

Land managers may seek to provide artificial tree hollows for one or more species likely to be impacted by a development. An estimated 15% of terrestrial vertebrate species in Australia, including threatened species, use hollows (Gibbons and Lindenmayer 2002). Different species generally have their own particular requirements for choosing suitable hollows (see Goldingay 2009). For the purpose of this book, we group and describe various artificial tree hollows as shown in Table 10.4 and outlined below.

Designs of artificial hollows are recommended by numerous sources, including published literature, businesses that sell ready-made artificial hollows, community groups, zoos and governments.

Artificial tree hollows for bats do not always need to be attached to a tree. They can be attached to buildings and other structures, even at the end of poles concreted into the ground. If artificial tree hollows are to be attached to living trees, the method used to attach them should not harm the tree or pose a safety hazard to anyone cutting the tree in the future. It is preferable to attach artificial hollows to trees using a loop or strap that is capable of expanding automatically or being expanded manually as the tree grows (e.g. Figure 10.9). If the type of strap chosen has the potential to cut into the bark, it should, ideally, be padded with rubber or similar material to prevent damaging the tree.

Table 10.4.   Types of artificial tree hollows

Type	Description
Artificial tree hollows for birds and non-flying mammals	A hollow box with a side hole for fauna to enter.
Artificial tree hollows for bats	A compartmentalised box with a hole on the lower side for hollow-dwelling bats to enter.
Artificial shedding bark	Materials fixed to a tree to imitate shedding bark and provide habitat for some hollow-dwelling bats.
Assisted hollow formation	Injuring trees to aid the formation of hollows.

Figure 10.9.   An artificial tree hollow suitable for some bird and mammal species erected in a tree at Osprey House Environmental Centre north of Brisbane, Queensland. Note the expandable loop connecting the artificial hollow to the trunk (Photo: D Gleeson)

Artificial tree hollows for birds and non-flying mammals

Artificial tree hollows constructed for birds and non-flying mammals are generally a rectangular hollow wooden box with a hole for fauna to enter (Figure 10.9). There are countless variations of this basic design, and they can be built from a range of materials. The specific dimensions of a nest box varies according to the species.

Birds and non-flying mammals use artificial tree hollows for nesting and roosting. These artificial tree hollows can also be used by tree frogs and reptiles. Detailed plans for various types of nest box designs are provided in Franks and Franks (2003).

Artificial tree hollows for bats

Bats that roost in tree hollows are small so the cavity design of the artificial tree hollows need not be as large as those for birds and non-flying mammals (Figure 10.10). Tuttle *et al.* (2004) provide various designs of bat roost boxes. One common design in Australia is a bat roost box constructed from multiple squares of exterior grade plywood fastened together with gaps (approximately 2 cm) that provide chambers in which microbats can roost (Figure 10.10b). The base is left open for bats to enter. This design also has an extended backing sheet of ply that provides bats with a landing area.

Artificial shedding bark

Some bats, such as the greater long-eared bat (*Nyctophilus timoriensis*), are known to roost under shedding bark. Tuttle *et al.* (2004) described how corrugated sheet metal wrapped around trees to mimic bark (fixed at the top and flared out at the bottom) could be used to supplement roost habitat for these bat species.

Figure 10.10. Open bottom bat boxes erected in a tree at (a) Osprey House Environmental Centre and (b) Dowse Lagoon, both north of Brisbane, Queensland (Photos: D Gleeson)

Assisted hollow formation

Natural hollows form after an aged tree has been subjected to physical injury from natural forces (e.g. rain, wind or fire). Fungi, bacteria and termites assist in decay of the heartwood (i.e. the inner core), thus forming a tree hollow. By understanding the way in which hollows form naturally, we can start to look at ways to artificially assist the formation of more hollows.

There have been various ways suggested to accelerate hollow formation. Adkins (2006) reviewed the feasibility and practicality of using fire to accelerate hollow formation in wood-production forests in Australia, and suggested that it is a potentially useful tool. Overseas, others have described making a cavity by drilling holes into the tree (Carey and Sanderson 1981), inoculating decay-causing fungi (Filip *et al.* 2004), applying pheromones to attract decay-causing beetles (Bull and Partridge 1986), deliberately killing the tree (e.g. by ecological thinning) and cutting the branches off the tree trunk (Adkins 2006).

Effectiveness

A growing number of studies are demonstrating that the use of artificial tree hollows can be an effective management tool for conserving particular hollow-dwelling fauna (e.g. Keeley and Tuttle 1999; Tuttle *et al.* 2004; Mooney and Pedler 2005; Flaquer *et al.* 2006; Beyer and Goldingay 2006; Durant *et al.* 2009). Although artificial tree hollows may have a limited life span (e.g. 10 years, depending on environmental conditions), a single artificial tree hollow has the potential to enable a breeding pair of birds to raise 10 clutches of offspring or house a whole bat colony. However, the potential usefulness of artificial hollows should not be used as a justification for removing hollow-bearing trees. The use of artificial tree hollows to accommodate a particular species should be viewed as largely experimental until a particular design has been shown to be effective for the species.

Artificial tree hollows can be installed any time of year. The time it takes for animals to occupy artificial tree hollows is variable and is probably influenced by factors such as time until the breeding season for the target species and population size of the target species. Some studies have shown that established artificial hollows are more likely to be occupied than more recently installed hollows (e.g. Durant *et al.* 2009). Even after sufficient time, 100% occupancy rates are unlikely. For example, artificial hollows have been used to assist in the recovery of the threatened Kangaroo Island glossy black-cockatoo (*Calyptorhynchus lathami halmaturinus*) and from over 80 artificial hollows (mainly made from PVC plumbing pipe) installed on Kangaroo Island at least half have been used for nesting by this species (Mooney and Pedler 2005). Of particular note, the artificial hollows in this study had a similar success rate to natural hollows (approximately 50%) (Mooney and Pedler 2005).

Artificial tree hollows that are not inhabited for an extended period may need to be moved to a more suitable location.

Faunal use of a well-designed artificial tree hollow is likely to reflect the relationship between the availability of natural hollows and the population size of the target species. If there are already plenty of hollows in an area, additional artificial hollows may not be needed. In many, but not all, cases, artificial tree hollows are more likely to be used when hollow availability is limited. Durant *et al.* (2009) found that sugar gliders were more likely to occupy artificial hollows when the density of trees bearing natural hollows was lower nearby.

Overseas, Tuttle *et al.* (2004) described how artificial shedding bark made from corrugated sheet metal has proved highly successful for the little brown bat (*Myotis lucifugus*) – a common bat in North America. Although not widely used in Australia, this is a relatively low-cost measure that could be used experimentally. Various materials, such as metal, plastic or fibreglass, that may be able to outlast natural shedding bark could be trialled (Tuttle *et al.* 2004). Reptiles and invertebrates are also likely to shelter under artificial shedding bark.

A number of authors have indicated that assisted hollow formation could be successful in Australia (Gibbons and Lindenmayer 2002; Brennan *et al.* 2005; Adkins 2006). Unfortunately, research into this topic is limited (Goldingay 2009) most likely due to the time it takes to achieve results from experimentation.

Design

Species generally have their own requirements when it comes to choosing a hollow (e.g. entrance size and shape, location on tree and depth). Very few scientific studies that offer animals a choice between designs of artificial tree hollows (i.e. preference studies) have been published for Australian birds and bat species (see comprehensive review by Goldingay and Stevens (2009)). Given this lack of data, it may be prudent to consider providing a range of hollow designs so that at least some are successful. It is critical that the design of the artificial tree hollow suits the target species(s) or it could be rejected.

If the objective is to compensate the loss of a particular natural hollow due to development, the substitute artificial tree hollow should be designed with similar characteristics so that the same species are encouraged to use the replacement hollow. The replacement hollow does not need to be an artificial hollow: a natural hollow-bearing tree planned to be removed could be salvaged and re-used instead (as discussed later in this chapter).

Species generally show a preference for artificial tree hollows with an entrance hole just wide enough to enter (Beyer and Goldingay 2006; Goldingay and Stevens 2009). Tight entrance sizes could be preferred because it would exclude larger predatory species and help to reduce competition for hollows by making them unsuitable for larger species.

Premature decay of the artificial hollow and the probability of its collapsing can be reduced by ensuring that high-quality materials have been used in its construction. Stainless steel and galvanised fixings provide rust resistance, while hardwood, polyvinyl chloride (PVC) or plastic hollows will last longer than softwoods such as pine. Applying a non-toxic paint to the exterior of wooden artificial hollows will increase its lifespan. Although artificial hollows should be designed and positioned to minimise the entry of rain, drainage holes in the base of an artificial hollow will help drain water and prevent young from drowning.

When choosing or designing an artificial tree hollow for birds and non-flying mammals, particular consideration should be given to the hollow volume, size of the entrance hole, hollow depth below the entrance and wall thickness (Figure 10.11).

When choosing or designing an artificial tree hollow for bats, particular consideration should be given to designing chambers large enough to prevent overheating (Figure 10.12). Also, Bat Conservation International (2011) states that artificial tree hollow for bats mounted

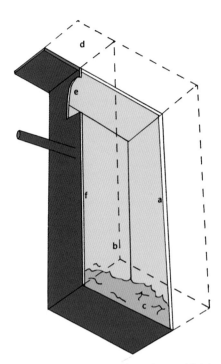

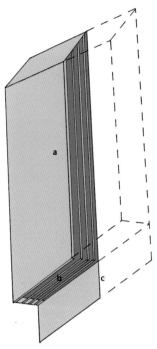

Figure 10.11.   A cross-section of an artificial tree hollow and the design elements that can affect its use by birds and arboreal fauna: (a) wall thickness provides insulation which can determine usage or nesting success (Trainor 1995); (b) suitable volume; (c) nesting material (e.g. sawdust) can be placed in the bottom of the box although this is optional as nesting individuals tend to bring their own material; (d) the roof of the box extends past the box entrance to buffer rain and sunlight; (e) fauna generally prefer an entrance hole just wide enough to enter (Goldingay and Stevens 2009), a bird perch is optional; (f) deeper hollows (larger depth of chamber between entrance and base) may provide security from predators (Saunders *et al.* 1982; Trainor 1995; Gibbons *et al.* 2002) and prevent nestlings from falling out (Saunders *et al.* 1982)

Figure 10.12.   A cross-section of an artificial tree hollow and design elements that can affect its use by bats: (a) inside: a surface suitable for bats to climb and cling on, e.g. grooves in wood (Irvine and Bender 1995), wire, suspended cloth; (b) the opening is located at the base as a narrow slit or no base at all to exclude or discourage non-target fauna (e.g. sugar gliders; Irvine and Bender 1995); (c) wall thickness provides insulation which can influence usage

on poles and buildings are more effective than those mounted on trees, perhaps due to unimpeded flight room.

Placement

The placement of an artificial hollow within the environment (i.e. in terms of topography, proximity to a food source, orientation, height and density) will affect its usage by fauna. Studies have shown that the topography of the landscape can influence usage for some species. For example, Durant *et al.* (2009) found that squirrel gliders primarily used artificial hollows installed in flat areas while sugar gliders mainly occupied artificial hollows located on mid-slopes and ridges. Both species of glider used artificial hollows placed in gullies.

The proximity of an artificial hollow to a food source may affect its usage by some species in some circumstances (e.g. Durant *et al.* 2009). For example, Rhodes and Jones (2011) found that bat boxes installed in urban Brisbane were more successful when within 1 km of grassed areas and within 5 km of forest remnants.

The orientation (i.e. aspect) of an artificial hollow can influence its microclimate (Ardia *et al.* 2006; Butler *et al.* 2009) and hatching success. Artificial tree hollows with side openings should be directed away from prevailing winds and shaded from direct sunlight. The orientation should favour protection from weather (Trainor 1995).

The density of hollows may affect their usage for some species. Durant *et al.* (2009) found that approximately three (as opposed to two) artificial hollows per 5 ha (the average home range) improves occupancy rates for sugar gliders. Similarly, Rhodes and Jones (2011) found that bat boxes installed in urban Brisbane were more successful when clustered in groups of at least six boxes. On the other hand, some species of cockatoo, for example, may only use artificial hollows placed further apart, because of behavioural interactions (see Saunders *et al.* 1982). Beyer and Goldingay (2006) describe how artificial tree hollows could prove to be important in assisting in the dispersal of species (e.g. squirrel glider), particularly through expanses of habitat or corridors without any hollows.

Due to a lack of information regarding placement requirements of artificial hollows for most species, an adaptive management approach will most likely be required. If an artificial hollow is not being used, it may need to be relocated to give it the best chance of benefiting the target species.

Most artificial tree hollow studies have installed the boxes between 3 and 8 m from the ground, but further resolution is needed on ideal height for particular species (see Beyer and Goldingay 2006). It is important that bat boxes are not installed above any obstructions (e.g. tree branches), so bats are able to drop before flight.

Maintenance

Ideally, artificial tree hollows should be designed to minimise maintenance requirements. Invasion by exotic and non-target native species is still possible, requiring regular monitoring and implementing measures to prevent undesirable species from invading. Published measures that have been used to successfully control invasion by such species are provided in Table 10.5. These control measures have had varying success, depending on environmental conditions. For example, Rhodes (2002a) found that frequent rain reduced the effect of ant preventative measures in bat boxes.

The recovery of the threatened glossy black-cockatoo population on Kangaroo Island has been partly attributed to the prevention of predation by possums on eggs and nestlings (Mooney and Pedler 2005). Possums were prevented from accessing trees bearing glossy black-cockatoo nests by pruning surrounding trees to prevent access via a continuous canopy and installing metal guards made of corrugated iron around the bases of trees to prevent access from the ground (Garnett *et al.* 1999).

Impacts on other aspects of the environment

Artificial tree hollows have the potential to increase populations of common or exotic species (Gibbons and Lindenmayer 2002). Invasion by exotic species (particularly common mynas and European honeybees) and non-target native species (e.g. corellas and common brushtail possums) can occur in artificial hollows just as they do in natural tree hollows. Invading fauna species may seek to use the artificial hollow as a roosting or nesting resource or they (e.g. pythons) may prey on the occupants already using the hollow.

Table 10.5. Measures that have been used to control invasion of artificial tree hollows by exotic and native non-target species

Invading species	Measure	Reference
Ants (general)	Talcum powder applied to the entrance and edges of the nest box to deter ants.	D. Harley (personal communication) cited in Gibbons and Lindenmayer 2002
	Talcum powder sprinkled inside of box incites ants to leave and lanolin grease around edges of the box prevents them from returning.	Franks and Franks 2003
	Ring of grease around trunk of smooth-skinned eucalypt encourages colony to leave the box.	Franks and Franks 2003
	Open bottom prevents ant infestations in bat boxes.	Rhodes 2002a
Wasps	2 cm roost spacing reduces wasp infestations in bat boxes.	Rhodes 2002a
European honeybee	Insecticide strip placed inside box kills bee colonies. This practice is hazardous.	Soderquist et al. 1996; Franks and Franks 2003
	Lining the ceiling of the nest box with carpet before installation may thwart attachment of wax comb to ceiling.	Soderquist et al. 1996
	A small box volume reduces incidents of hive building.	Goldingay et al. 2007
	Greasing the underside of the lid and top of the walls with marine grease or lanolin prevents bees from attaching honeycomb.	Franks and Franks 2003
	2 cm roost spacing reduces bee infestations in bat boxes.	Rhodes 2002b
Exotic birds (general)	A slit (narrow) entrance on bat boxes excludes exotic birds.	Smith and Agnew 2002
Common myna	A board of ply attached to overhanging box lid and positioned approximately 10 cm parallel to the front face (i.e. side including entrance hole) of the box successfully excluded the common myna, but not native species.	Homan 2000
	Nest removal deters nesting, but may need to be repeated several times	Franks and Franks 2003
Common starling	Starlings avoid nest boxes with painted white interiors	Lumsden 1989
Currawongs	A 10 cm length of 7.35 mm diameter PVC tube leading into centre of box to create a spout entrance stopped predation on crimson rosella chicks.	Krebs 1998
Galahs, little corellas	Deployment of decoy artificial hollows may alleviate competition.	Goldingay and Stevens 2009
Feathertail glider	Open-bottom roost box design for bats prevents feathertail gliders establishing a leaf nest.	Goldingay and Stevens 2009
Common brushtail possum	Metal guards (e.g. corrugated iron) around base of tree prevents possums from killing glossy black-cockatoo nestlings.	Garnett et al. 1999

Assisted hollow formation (e.g. making a cavity by drilling into the tree and inoculating decay causing fungi) can have adverse impacts on trees, potentially leading to tree mortality. Further research is needed to improve the effectiveness of this measure so that adverse impacts on trees are minimised.

Artificial subterranean bat roosts

Overview	
What:	Structures intended to provide habitat for cave-dwelling bats.
Where:	Useful where development (e.g. a mine) could destroy or introduce significant disturbance at a cave (or similar structure) used by cave-dwelling bats.
Pros:	Provides replacement or additional habitat where important roosts coincide with plans for development.
Cons:	Needs further testing. Creation of a purpose-built structure may require ground disturbance.
Cost:	Medium–high.
Companion measures:	Use of alternatives to barbed wire (Chapter 8); revegetation and restoration of ecosystems (this chapter).
Alternative measures:	Avoid disturbance to bat caves (Chapter 4).

Description and potential applications

A cave used by bats may need to be removed in order for a proposed development (e.g. a mine) to proceed. If alternative habitat is limited nearby, the removal of the cave may result in the loss of a bat colony and, in some cases, the local extinction of a bat species. Artificial subterranean bat roosts are structures intended to maintain or increase the habitat opportunities for cave-dwelling bats.

The concept of purposefully creating subterranean habitat to conserve cave-dwelling bats in Australia was described by Thomson (2002). It is not surprising that bats can use purposefully created subterranean habitat because almost one third of Australian bat species have been recorded using structures built for other purposes, either as maternity, hibernation and/or temporary roosts (Hall and Richards 1998; Thomson 2002; Sanderson *et al.* 2006; Law and Chidel 2007) (Figure 10.13).

There are several different types of artificial subterranean bat roosts, which can be split into two groups: habitat created by converting existing human-made structures (i.e. converted structures) and habitat created by purposefully building new structures (e.g. bat tunnels, bat culverts and bat domes). The various types of artificial subterranean bat roosts are described in Table 10.6 and below.

Converted structures

Existing structures built for human purposes (e.g. tunnels, stormwater drains, culverts, bridges and buildings) can be modified to increase habitat opportunities for bats. Examples of modifications include blocking excess holes in a structure to stabilise the microclimate (while still allowing at least two openings for air circulation), providing rough ceiling surfaces that bats can cling to and ensuring walls are smooth to minimise predator access to roosting bats (Mitchell-Jones 2004; Thomson 2002).

Figure 10.13.   Disused mines such as shown here are often used by bats (Photo: J Gleeson)

Drilling small holes (<10 cm) into the ceilings of concrete structures (e.g. road culverts) has been anecdotally shown to provide roost habitat for some bats, such as Troughton's cave bat (*Vespadelus troughtoni*) (Thomson 2002). Unfortunately, there is a lack of published information on the most efficient size and density of holes. However, this is a relatively low-cost modification that could be used experimentally. In other countries, such as Japan, purpose-built rough concrete substrates and wooden bat boxes have been installed on smooth concrete surfaces with success (Mukoyama 2005).

Bat tunnel

Artificial subterranean bat roosts have been unintentionally, but successfully, created as a by-product of historic underground mining in many parts of Australia and around the world. In fact, the need to conserve these disused mines is now widely recognised, particularly as some of the largest subterranean bat populations in Australia are now known to be using these mines (Hall *et al.* 1997; Armstrong and Anstee 2000; Armstrong 2001; Hutson *et al.* 2001).

Table 10.6.   Types of artificial subterranean bat roosts

Type	Description
Converted structures	Existing structures (e.g. tunnels, stormwater drains, culverts, bridges, buildings or disused mine adits) modified to provide habitat for some bat species.
Bat tunnel	Large holes tunnelled into the side of hills or rises to provide bat habitat (i.e. similar to historic mine adits – Figure 10.13).
Bat culvert	Drainage culverts designed to provide habitat for bats incorporated into the design of developments (e.g. roads).
Bat dome	A domed structure built above the ground to simulate subterranean habitat.

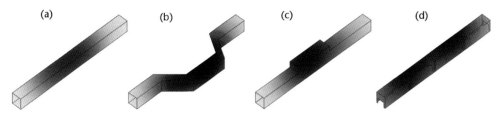

Figure 10.14. Various culvert designs to be used by bats: (a) standard culvert with minimal darkness; (b) culvert with bends; (c) culvert with middle dome; and (d) partially blocked culvert

Thomson (2002) described the optimal artificial subterranean bat habitat as an adit tunnelled into hard stable rock, with higher and lower ceilings to create multiple microclimates. Horizontal mine adits are thought to be better than vertical shafts because they provide a variety of microclimates (New South Wales National Parks and Wildlife Service 2001a). Bat tunnels need to be engineered to be structurally sound and able to withstand environment conditions (e.g. extremes in weather or seismic waves resulting from mine blasts), as well as changes in watertable levels.

Bat tunnels can be incorporated into the rehabilitation of some post-mine landforms (Hall and Richards 1998; Ducummon 1999; Thomson 2002). Open cut mining involves the removal of earth to access mineral ore and often the replacement of the overburden into the mined-out pit. Thomson (2002) described how a bat tunnel made from recycled mine tyres was incorporated into the rehabilitation of an Australian mine. This method of creating a bat tunnel has achieved success overseas. For example, buried culverts and portable buildings have been incorporated into mine rehabilitation to create bat tunnels in Idaho and Nevada, USA (Ducummon 1999). The post-mine landforms should be stable for the longevity of the artificial subterranean bat roost.

Bat culvert

Some bat species, such as the threatened little bent-winged bat (*Miniopterus australis*), have been observed using standard drainage culverts incorporated in the design of roads (Churchill 2008). The design of standard drainage culverts as shown in Figure 10.14a can be modified in at least three ways to provide better habitat opportunities for bats.

Firstly, culverts can be designed with the middle portion offset to create bends and a dark roosting area in the middle portion of the culvert (Queensland Department of Transport and Main Roads 2010) (Figure 10.14b). A possible downside to this design is that air can pass freely through the culvert as wind, which may deter some bat species. Secondly, adding a dome (~0.6 m higher than the majority of the culvert in the middle portion of a typical culvert) can trap warm air, reduce air movement and reduce light, providing more suitable conditions for bats (Keeley and Tuttle 1999) (Figure 10.14c). Thirdly, bat habitat can be improved by partially blocking the ends of the tunnel to stabilise air movement and/or partially blocking sections of the tunnel to trap warm air (Mitchell-Jones 2004) (Figure 10.14d). These modifications would also reduce daylight in the tunnel.

In the UK, approximately 24 artificial subterranean bat roosts were known to have been constructed as of 2004, with the design of many of these structures based on concrete tunnels (Mitchell-Jones 2004).

Bat dome

A bat dome was constructed at the Selah Bamberger Ranch Preserve, Texas, USA in 1998 (Sander 1997; Bamberger Ranch Preserve 2011). The domed structure (covering an area of approximately 280 m²) was built at ground level from steel, metal and concrete and was covered with soil and revegetated to blend into the surroundings (Sander 1997). Additional roosting substrates were provided inside the artificial cave by hanging panels (e.g. wood) from the ceilings (Sander 1997).

In Japan, several species have used specially constructed wooden bat houses, either as day roosts or as a night resting refuge. These structures were purpose built for species that needed to be shifted from buildings and temples after becoming a nuisance or causing damage, and have design features that encourage bat habitation and allow guano build-up to be removed easily. Some are simply wooden buildings, while others are old railway carriages that have been covered in earth (Mukoyama 2005).

Effectiveness

Thomson (2002) described how a bat tunnel made from recycled mine tyres in an Australian mine was colonised by bats, but the ongoing success of this roost is not known. The use of artificial subterranean bat roosts to accommodate a particular bat species should be viewed as largely experimental until repeatable design specifications for successful roosts are known.

In Europe, structures built for other purposes, such as abandoned railway tunnels and underground bunkers, have been successfully modified to provide suitable conditions for bats to roost (Hutson *et al.* 2001; Mitchell-Jones 2004). The bat dome in Bamberger Ranch Preserve, Texas mentioned above was not used by bats for a considerable number of years, but it is now used by an estimated 200 000 Mexican free-tailed bats (*Tadarida brasiliensis*) (Bamberger Ranch Preserve 2011).

Factors likely to influence the effectiveness of artificial subterranean bat roosts include the design and placement of the roost, as well as consideration of species-specific traits of the bats for which they are constructed. These factors are discussed below.

Design

Different bat species have differing roost requirements, according to their physiological, reproductive and ecological needs. For example, the pregnant females of some species, such as ghost bats (*Macroderma gigas*), aggregate into maternity colonies with relatively warm and humid microclimates to give birth and suckle young; others, such as the orange leaf-nosed bat (*Rhinonicteris aurantia*) in the Northern Territory (Churchill 1995), simply aggregate to mate before dispersing to other roosts; some species, such as the orange leaf-nosed bat (Churchill 1991; Baudinette *et al.* 2000; Armstrong 2001), must have warm and humid microclimates to conserve their water and heat balance; other species, such as the southern bent-wing bat (*Miniopterus schreibersii bassani*) (Dwyer 1969; Codd *et al.* 2003), choose cooler microclimates to hibernate in large over-wintering aggregations, and other cave roosting species may use several roosts as transient day and night roosts (Lewis 1995; Hall and Richards 1998; Ramos Pereira 2009). The design of artificial roosts therefore must consider the needs of the species that are to occupy them.

Generally, the following design elements will increase the effectiveness of an artificial subterranean bat roost (Thomson 2002):

- sufficient airflow to replace air (but not to create wind)
- multiple ceiling heights that can trap warm, humid pockets of air
- darkness
- sufficient buffering of outside extremes in humidity and temperature
- roosting substrate (rough ceiling)
- clean roost surfaces (i.e. no heavy dust).

If predators are able to enter the structure, the ceiling of the structure should be high enough to prevent predators from reaching bats roosting on the ceiling, and smooth walls would also inhibit predators from climbing walls. If predators are likely to be an issue, an appropriately designed steel grill gate (e.g. Thomson 2002) may need to be considered. Such a gate may also deter human entry, as would an information sign.

It may take a while for bats to move into an artificial subterranean roost of their own accord (Mitchell-Jones 2004; Bamberger Ranch Preserve 2011). There have been several suggestions on how to encourage bats to find and use newly created artificial structures. Hall and Richards (1998) suggested that artificial subterranean bat roosts could be littered with guano to encourage bat occupation, but this is yet to be tested. Simple manual relocation may not have high initial success, but experience overseas has shown that some species that form colonies in buildings can eventually be coaxed into a new home in this manner. The level of disturbance should be considered in any plan, because some species are very sensitive to intrusions and handling. A carefully designed program of exclusion from the original roost should accompany efforts to encourage habitation in a newly created alternative, and sufficient time should be allowed for this when planning future developments such as mines.

It is important to consider how long a proposed structure is likely to be in place, particularly for high conservation significant species, where the species is dependent on the longevity of the structure. Longevity will vary with the design of the artificial subterranean roost and the environmental pressures acting on it.

Placement

Site selection is an important step in creating an effective subterranean bat habitat. The artificial subterranean bat roost would need to be located within the range of the target bat species, in an area containing access to food and near a network of cave types (maternity and temporary roosts). Ideally, the site would be free from major predators, and in an area that will not be visited frequently by people. Preventing human entry could be achieved by having the entrance above head-height or removing earth from in front of the entrance, if it cannot be raised. Installing the structures on a slope can enhance the ability of the structure to trap humidity in roosting chambers positioned upslope, but the gradient needs to be sufficiently gentle to avoid erosion of the earth covering. In addition, the cost of installing culverts or other infrastructure can be considerable and some thought will obviously need to be given to access for heavy vehicles, especially in remote areas or rough terrain.

Species-specific traits

Bats have specific microclimatic requirements for roost sites, which vary between species and season (Baudinette 1994; Churchill 2008), so designing subterranean bat habitat for a particular bat species may prove challenging. Depending on the species, the caves used by bats range from completely dark, isothermal, hot, high humidity conditions to draughty, semi-illuminated sites with low humidity (Hall and Richards 1998). High humidity may be the hardest aspect to re-create, but may be achieved by extending an adit below the watertable (Hall and Richards 1998), creating domes, including pools and positioning the structure on an incline.

If the bat species is known to occupy disused mines or other artificial structures, the species may be more likely to use a purpose-built structure. If the aim is to create a roost for a particular bat species, it is best to mimic the subterranean bat habitat known to be used by that species. For example, Hall and Richards (1998) describe how some species, such as the ghost bat (*Macroderma gigas*), prefer multiple entrances, while others prefer a single adit, such as bent-winged bats (*Miniopterus* spp.).

Impacts on other aspects of the environment

Some bats are carnivorous, such as the ghost bat (*Macroderma gigas*) and greater broad-nosed bat (*Scoteanax rueppellii*) (Churchill 2008). The use of an artificial subterranean roost by a carnivorous bat species may result in the predation upon, or displacement of, other bat species in the feeding range of the carnivorous bat. In this situation, site selection is particularly important, even more so if the potential prey is a threatened species. Design features can also be incorporated that partition the species inside, such as providing separate roosting areas.

In situations where the creation of a suitable bat roost requires land clearance, the negative impacts from land clearance should be weighed up against the possible gains to be had from creating bat habitat. Alternatively, land clearance can be avoided or minimised by converting existing structures to provide bat habitat, incorporating bat habitat into mines landforms as part of their rehabilitation or incorporating bat culverts into developments already requiring land clearance (e.g. under roads).

Companion measures that may be suitable to further reduce the impact of development on cave-dwelling bats include using alternatives to barbed wire (Chapter 8) or enhancing foraging resources by revegetating and restoring ecosystems (this chapter).

Artificial nesting platforms

Overview	
What:	Structures that are intended to provide birds, typically birds of prey, with a roosting and nesting substrate.
Where:	Where development has reduced, or will reduce, the number of similar natural perches or nesting sites available to the avian species that use them.
Pros:	Encourages birds to remain in an area or return to an area previously affected by development.
Cons:	May change prevailing predator–prey relationships and lead to a higher risk of predation for prey species.
Cost:	Low.
Companion measures:	Revegetation and restoration of ecosystems (this chapter).
Alternative measures:	Avoid disturbance to natural perches or nesting sites (Chapter 4); salvaged habitat features (this chapter).

Description and application

Artificial nesting platforms are structures that are intended to provide birds, typically birds of prey, with a roosting and nesting substrate. These birds would normally nest in tall isolated trees. Artificial nesting platforms can be an alternative habitat resource if development has reduced, or will reduce, the number of similar natural sites available to the avian species that use them.

Suitable natural perches or nesting sites can be limited in some situations. It is not uncommon for birds to choose to nest on structures made for other purposes, such as electrical transmission line towers, the edges of high-rise buildings and bridge structures. This of course is not always ideal for the bird and can be undesirable for the development.

Providing purpose-built artificial nesting platforms and perches, especially for birds of prey, is becoming commonplace around the world. Artificial nesting platforms built for

Figure 10.15.   Pair of ospreys using an artificial nesting platform erected near the Harwood Bridge, Northern Rivers region of New South Wales (Photos: R Whitney)

ospreys (*Pandion haliaetus*) (Box 10.4) and bald eagles (*Haliaeetus leucocephalus*) have been used at least since the early 1960s, such as those in the Chesapeake Bay area of the USA (Reese 1970).

Figure 10.15 shows an artificial nesting platform atop a pole with lower perches that was erected near the Harwood Bridge, Northern Rivers region of New South Wales. A pair of ospreys were quick to accept the artificial nesting platform (Box. 10.4). This nesting platform was made from a 30 m tall wooden pole (New South Wales Government Roads and Traffic Authority 2009).

In this chapter, we describe purpose-built tall structures with perches and nesting platforms at the top. There are other ways of providing roosting resources for different types of birds, such as the installation of perches within or around a wildlife-friendly dam and the use of salvaged dead trees (which are both discussed elsewhere in this chapter). Existing developments can also be modified to provide more suitable nesting and roosting conditions (e.g. adding structures to existing buildings).

Effectiveness

Practical experience is demonstrating that a well-placed artificial nesting platform can be very effective for particular birds (e.g. Box 10.4). In fact, in America, a long-term study has shown that ospreys using platforms made by people had a higher nesting success rate (62.9%) compared with nests built by ospreys in trees (45.9%) (Houston and Scott 1992). Although there are successes, further research is needed to test the effectiveness of artificial nesting platforms for different bird species in Australia.

Manipulative field experiments overseas have revealed that adding artificial perches can increase visitation by some raptor species, such as American kestrels (*Falco sparverius*), but not others, such as the northern harrier (*Circus cyaneus*) (Wolff *et al.* 1999; Kim *et al.* 2003).

Box 10.4. Ospreys and artificial nesting platforms in Australia

Ospreys (*Pandion haliaetus*) are cosmopolitan, fish-hunting raptors. Ospreys usually favour tall trees with an unobstructed view of the surrounding area for nesting. Some of their nests are used for up to 70 years and are renovated each season. In Australia, their nesting habitat has been reduced by encroaching urban development near waterways and logging of tall trees.

Artificial nesting platforms have been erected to supplement existing habitat that has been diminished by encroaching urban development. An artificial nesting platform purposely built for raptors at Osprey House beside the Pine River Estuary near the city of Brisbane was first occupied approximately 2.5 years after construction by a pair of ospreys, which then raised one offspring.

There are also documented examples of wildlife managers successfully encouraging ospreys to abandon nests on existing infrastructure or on trees earmarked to be removed for development in favour of using artificial nesting platforms. For example, ospreys nesting on the Harwood Bridge structure in New South Wales were successfully encouraged to nest instead on an artificial nesting platform (a nest cradle fixed to the top of a 30 m pole) purposely erected nearby (New South Wales Government Roads and Traffic Authority 2009) (Figure 10.15).

Design

The effectiveness of a nesting platform may be affected by height and quality of the structure. The most effective design of a nesting platform will vary between species. Yet, little or no research has been undertaken to determine the most appropriate design. Instead, many designs were probably developed to mimic natural nesting platforms. Increased height of the platform may improve hunting success for predatory birds (Andersson *et al.* 2009).

Placement

The effectiveness of nesting platforms may depend on the scarcity of nesting opportunities, as well as the local abundance of the target species. The success of the artificial nesting platforms erected at Martin National Wildlife Refuge, USA was attributed to a lack of suitable nest sites and the sizeable population of ospreys that was present (Rhodes 1972). The spacing between artificial nesting platforms erected at the Refuge (varying from less than 180 m to more than 800 m) did not correlate with osprey use of the platforms or production of chicks (Rhodes 1972).

Impacts on other aspects of the environment

The addition of nesting platforms for birds of prey may change prevailing predator–prey relationships and lead to a higher risk of predation for prey species (Andersson *et al.* 2009). The area of suitable habitat for prey species may be reduced as a result. This is especially problematic if the prey species likely to be affected are threatened species.

Materials from sustainable practices should be preferred in the construction of artificial nesting platforms. Trees that are removed from a development site can be salvaged and re-used (as discussed later in this chapter).

Artificial retreats for reptiles

Overview	
What:	Rock substitutes and purposely created burrows intended to provide refuge habitat for reptiles.
Where:	Where a development has resulted, or will result in, a reduction in the availability of natural retreats for reptiles and natural retreats are lacking.
Pros:	Encourages reptiles to remain in an area or return to an area previously affected by development.
Cons:	May attract undesirable predators, competitors or exotic animals.
Cost:	Low.
Companion measures:	Natural habitat linkages (Chapter 8); revegetation and restoration of ecosystems (this chapter).
Alternative measures:	Avoid disturbance to natural retreats (Chapter 4); salvaged habitat features (this chapter).

Description and application

Artificial retreats or refuges for reptiles can be created from materials that substitute their natural retreats (e.g. artificial rocks, concrete pavers or corrugated iron sheets), or through purposely created burrows. Reptiles use retreats for protection against predators and buffering of environmental conditions, such as thermal extremes. Artificial retreats are used if a development results in a reduction in the availability of natural retreats. Re-use of salvaged habitat features from development sites (e.g. rocks and logs) for reptile habitat is discussed in the next section.

Although the focus of this section is on artificial retreats suitable for reptile use, it should be noted that other animal groups use retreats. For example, artificial tree hollows are used as retreats by many bird and mammal species as discussed earlier in this chapter.

It is not surprising that reptiles can use purposefully created retreats, because structures made by people for other purposes are regularly used by reptiles as retreat sites. In fact, employing artificial shelters is a recognised technique to survey some reptiles, such as striped legless lizards (*Delma impar*) (Department of Sustainability, Environment, Water, Population and Conservation 2011d). Blue-tongued lizards (*Tiliqua scincoides*), for example, readily use drainage pipes, sheds and crevices in concrete for shelter (Koenig *et al.* 2001).

Natural formations of loose rocks suitable for animals to shelter beneath takes hundreds of thousands of years, so they should be considered to all intents and purposes non-renewable resources (Croak *et al.* 2010). This type of natural habitat is best protected rather than attempting to reinstate habitat with artificial substitutes later. Natural rocks have the advantage of looking authentic and such amenity is particularly important in reserves.

A spate of interesting investigations in recent years underlines the promising potential of artificial retreats for use as restoration tools for reptile communities. The provision of artificial retreats is likely to speed up the recolonisation of disturbed areas (e.g. rehabilitating mine sites) by providing essential habitat for key species (e.g. Marquez-Ferrando *et al.* 2009). Artificial retreats could also be used to assist re-introductions by encouraging translocated individuals to remain close to their point of release (Arida and Bull 2008). Translocation is discussed further in Chapter 11.

Supplementation of habitat with artificial retreats is a particularly promising way of retaining or attracting reptile species that rely on specific habitat features (i.e. habitat specialists).

Table 10.7.   Examples of different types of artificial retreats for reptiles

Type of artificial retreat	Description
Artificial rocks	Custom made from fibre-reinforced cement to resemble and mimic the properties of natural rocks.
Concrete pavers or tiles	Pavers slightly raised from the ground with spacers attached at the corners.
Corrugated iron or onduline sheets	Triple-layered stacks of corrugated iron or onduline (a corrugated roofing product made from organic fibres saturated with bitumen) sheets separated by spacers.
Earth burrow	Can be made simply by hammering a steel rod into the ground or more elaborately, for example, by using a wooden box connected to the surface via two PVC ringed pipes.

Threatened habitat specialists that are known to make use of artificial retreats include the broad-headed snake (*Hoplocephalus bungaroides*), which typically uses sandstone rocks as diurnal retreat sites (Webb and Shine 2000), and the pygmy blue tongue lizard (*Tiliqua adelaidensis*), which is dependent on burrows (Souter *et al.* 2004).

There are several types of artificial retreats for reptiles and these can be split into two groups: materials that substitute natural retreats or purposely created burrows (Table 10.7).

Effectiveness

Recent research has shown that artificial retreats have the potential to be very effective in reinstating refuge habitat for reptiles, including the threatened broad-headed snake (Croak *et al.* 2010) and the threatened pygmy blue tongue lizard (Souter *et al.* 2004). All studies reviewed indicated that artificial retreats were used to some extent by at least one species of reptile.

Artificial rocks have been shown to be used by reptiles on sandstone plateaus south of Sydney. Artificial rocks that had similar thermal regimes and crevice structures to natural rocks set on sandstone plateaus were effective (Croak *et al.* 2010). Most (81.8%) of the artificial rocks were colonised by reptiles after 40 weeks. Six lizard and two snake species, one of which was the broad-headed snake, used the artificial rocks (Croak *et al.* 2010).

Square, grey concrete pavers (19 cm wide × 5 cm thick) were used as an artificial retreat by velvet geckos, the prey of the broad-headed snake, south of Sydney (Webb and Shine 2000). Concrete roofing tiles (39 × 32 × 2.5 cm) were used by lizards during a field experiment undertaken in a coastal shrubland remnant in New Zealand (Lettink and Cree 2007).

In New Zealand, triple-layered corrugated iron sheets were used by terrestrial lizards during a field experiment in a coastal shrubland remnant (Lettink and Cree 2007). Lettink and Cree (2007) did find, however, that common geckos (*Hoplodactylus maculatus*) in New Zealand strongly preferred multi-layered Onduline stacks over triple-layered corrugated iron sheets and concrete roofing tiles. After conducting field tests and lizard preference tests in the laboratory, Thierry *et al.* (2009) recommended triple-layered stacks of corrugated Onduline sheets separated by spacers for restoration projects. A range of thermal microclimates were created by the three layers (Thierry *et al.* 2009).

Supplementation with artificial burrows made by hammering a steel rod (1.3 or 1.7 cm diameter) 30 cm into the ground was associated with an increase in local population density of the pygmy blue tongue lizard in South Australia (Souter *et al.* 2004). Similarly, the provision of artificial burrows on an island off the French Atlantic coast encouraged a threatened lizard (*Timon lepidus*) to swiftly recolonise areas once occupied (Souter *et al.* 2004).

Species generally have their own requirements when selecting a retreat site. The characters likely to influence a species' preference are its thermal properties, substrate type and security from predation (Arida and Bull 2008). Some studies have found that preferred retreats tend to have narrow, single entrances and deeper cavities (Arida and Bull 2008). It is likely that members of the same species will have varying requirements based on intra-specific differences such as body size.

If a particular species or faunal group is to be targeted, it is critical that the design of the artificial retreat suits the target species(s) or it could be rejected and deemed ineffective. For restoration projects, it can be beneficial to provide a range of artificial retreats that encompass varying entrance sizes and internal dimensions so that a broad range of species and size classes can be accommodated.

The effectiveness of artificial retreats may depend on the scarcity of natural retreat sites as well as the abundance of the target species occurring in the area.

Impacts on other aspects of the environment

The addition of artificial retreats can attract predators that may prey on the target species. Souter *et al.* (2004) found that additional artificial burrows corresponded with an increase in the number of centipedes, which could potentially prey on the reptiles.

Poorly designed artificial retreats can become a haven for rats and other small exotic animals. Resident snakes may help reduce exotic animals numbers, but otherwise a baiting program would be required to control exotic animals using artificial retreats.

Salvaged habitat features

Overview	
What:	Natural features such as hollow logs, solid logs, tree hollows, stag trees and rocks salvaged from an area earmarked for development and used as habitat for fauna.
Where:	Can be salvaged from areas undergoing clearance for development.
Pros:	Adding habitat features can encourage fauna to stay in an area or assist in the earlier recolonisation of previously disturbed areas.
Cons:	May attract undesirable predators, competitors or exotic animals.
Cost:	Low–medium.
Companion measures:	Revegetation and restoration of ecosystems (this chapter) as well as on land bridges and fauna underpasses (Chapter 8).
Alternative measures:	Artificial tree hollows; artificial nesting platforms; artificial retreats for reptiles (all this chapter).

Description and application

Habitat features such as hollow logs, solid logs, tree hollows, stag trees and rocks are often present in areas undergoing development. Instead of destroying these habitat features, they can be removed before land clearance and relocated to provide fauna habitat elsewhere.

Tree hollows and hollow logs in particular are well worth salvaging because lengthy periods of time are generally required for them to develop. For example, eucalypts are generally between 120 and 180 years old before they develop hollows suitable for vertebrate fauna (Gibbons and Lindenmayer 2002). Salvaged tree hollows can either be re-used as hollows in trees or as hollow logs resting on the ground. Natural formations of loose rocks suitable for animals to shelter beneath takes hundreds of thousands of years, so there is great merit in salvaging what would otherwise be lost to development.

Table 10.8.   Types of salvaged habitat features and examples of Australian native animals that could potentially use them

Type	Description
Tree hollows	Hollows formed naturally in trees (e.g. hollow branches or trunks) can be salvaged and re-fitted into a tree for tree frogs, birds, hollow dwelling mammals including bats or laid on the ground for invertebrates, reptiles, echidnas, quolls, antechinuses, numbats and bandicoots.
Salvaged wood	Any wood from a tree including logs, branches and roots can be laid on the ground for invertebrates and reptiles, among other species.
Timber fence posts	Old timber fence posts that are no longer useful can be laid on the ground for some frogs, geckos, lizards, skinks, snakes and the fat-tailed dunnart.
Stags	Trees can be transplanted dead or alive in an upright position so that they become stags (i.e. standing dead trees) for birds (including raptors and owls) and some hollow-dwelling mammals, including bats, if the tree possesses hollows.
Rocks	Rocks carefully positioned, for example, to form crevices, may be used by insects, spiders, geckos, skinks and snakes.

Salvaged habitat features are often used in areas undergoing rehabilitation (e.g. degraded areas in general, as well as rehabilitating mine sites). Structural habitat heterogeneity affects biodiversity, so habitats that have been degraded might be enhanced by re-establishing important habitat features. Adding habitat features can assist early recolonisation by fauna.

Salvaged habitat features can be strategically positioned in linkages between habitat areas such as wildlife corridors, land bridges and fauna underpasses. Providing habitat in this way encourages animals to use the linkage. Piled rocks and hollow logs, for example, can provide shelter and opportunities to escape predators for wildlife using the linkage. Examples of Australian animals likely to use these types of salvaged habitat features are provided in Table 10.8.

Tree hollows

Natural tree hollows (e.g. hollow branches) that have been salvaged from a development site will often require modifications before they are ready to be re-used. Natural hollows intended for use by birds and arboreal mammals should be cut into lengths of at least 0.5 m (Trainor 1995) so that there can be a reasonable depth of chamber between the entrance and the base. Deeper hollows are thought to provide security from predators (Saunders *et al.* 1982; Gibbons *et al.* 2002). If the bottom of the hollow is open after being cut to size, it can be blocked with salvaged timber such as a small length of log. The use of materials that inhibit drainage can lead to nest failure and should be avoided. Natural tree hollows can also be salvaged and re-used as hollow logs resting on the ground. This is discussed further in the next section. Tree hollows containing native stingless bees (e.g. *Trigona* spp.) are also sometimes salvaged and relocated. Native bees can be easy to salvage, because they are stingless, and will continue to thrive if relocated to suitable habitat.

Salvaged wood

Fallen wood (also known as coarse woody debris or habitat logs) contributes to the habitat structural complexity of an ecosystem. Fallen wood can be produced naturally owing to the shedding of branches from living trees (e.g. many species of eucalypts) right through to larger scale event such as a whole tree collapsing during high winds or after death. All parts of the tree to be cleared for development, from the trunk to the smallest branches and roots, can be salvaged and used to mimic naturally fallen wood. By providing an assortment of tree parts in a range of state of decay, more species are likely to benefit.

Fallen wood in natural Australian forests can benefit flora and fauna in many ways, as detailed in a notable review by Lindenmayer *et al.* (2002). Logs will often have crevices (especially in log ends), shedding bark and hollows suitable to be used as shelter for a range of invertebrate and vertebrate fauna. Besides shelter, logs can be used for nesting, hibernation and basking (reptiles), and as a foraging substrate (some snakes and birds). The brown treecreeper (*Climacteris picumnus*) uses fallen wood as a foraging substrate (Mac Nally *et al.* 2002).

Fallen wood can be used: to aid faunal movement (e.g. logs are used as runways by rodents and possums); as sites for social behaviours (e.g. territorial scat deposition by the common wombat and mountain brushtail possum); and to act as germination sites for plants and microhabitat for fungi. Fallen wood is used by a large range of invertebrates, which, apart from being significant in their own right, provide food for vertebrates.

It is likely that salvaged wood can provide the same types of benefits to fauna when applied to a rehabilitating site or a wildlife linkage (Figure 10.16). Salvaged logs and branches can be scattered or they can be piled together along with rocks and soil to provide dens. Such dens are provided at a density of three to five per hectare across rehabilitating mine pits at the Boddington Bauxite Mine in Western Australia to provide habitat for the threatened chuditch (*Dasyurus geoffroii*) (Brennan *et al.* 2005). These dens would also be expected to provide shelter for other animals crossing the bare mine pits.

Timber fence posts

Not all habitat features suitable for salvaging are in their natural state. Old timber fence posts can be recycled and used as fauna habitat. Salvaged timber fence posts are likely to provide similar benefits to wildlife as the larger types of fallen wood (i.e. logs).

Figure 10.16   Logs (some with hollows), piled rocks and scattered branches have been provided as habitat for fauna on a rehabilitating mine site, northern New South Wales (Photo: D Gleeson)

Figure 10.17.   A tree trunk transplanted upright to become a stag (Photo: P Smith)

Stags

Transplanting trees (dead or alive) upright so that they become stags (i.e. standing dead trees; also known as snags in some countries) can potentially enhance habitat for a range of fauna species (Figure 10.17). Transplanted stags could be used as perches by birds, refuges by arboreal mammals (e.g. to temporarily escape predators in bare areas where there are no other trees) and, if the stag has hollows, crevices or exfoliating bark, it can provide roosting or breeding opportunities for a range of fauna. Transplanted stags that are used as perches by birds might also help revegetation efforts and/or increase flora biodiversity because the birds' droppings are likely to contain plant seeds.

Rocks

Rocks can be salvaged from a development site and carefully positioned so that they are likely to provide habitat for fauna in another area (Figure 10.18). The shape of the rocks and the nature of placement are likely to dictate whether the salvaged rock will provide habitat and, if so, for which species. Crevice size (e.g. the space under the positioned rock) and temperature, for example, affect whether reptiles will use a retreat site (Pike *et al.* 2010).

Effectiveness

From a practical perspective, natural tree hollows that have been salvaged from trees are often very heavy and oddly shaped, making it difficult to secure them to trees or poles. However, they can be less vulnerable to chewing by parrots than the artificial alternatives.

Most demonstrations of the importance of fallen wood in influencing species occurrence have been correlative, but recent experimental manipulations of salvaged wood loads have found that the yellow-footed antechinus (*Antechinus flavipes*) prefers sites with more salvaged

Figure 10.18.   Salvaged habitat features such as hollow logs and rocks can provide shelter for fauna (Photos: D Gleeson)

wood, particularly for breeding (Mac Nally and Horrocks 2007, 2008). Higher loads of salvaged wood derived from tree crowns have also been shown to increase the species richness and numbers of birds (Mac Nally and Horrocks 2007, 2008).

A study undertaken in Victoria revealed that placing salvaged fence posts in grassy landscapes is effective as a potential technique for habitat restoration and fauna conservation (Michael *et al.* 2004). The recycled fence posts were quickly used by many terrestrial fauna, including species that are considered rare.

Trees transplanted onto mined sites to act as stags have been shown to be effective in attracting birds. McClanahan and Wolfe (1993) showed that stags transplanted onto a mined site in Florida attracted birds and, in addition, they found that there was a higher abundance and diversity of bird-dispersed plants under stags. Salvaging trees to use as stags has undergone limited trials at mine sites in Australia. Trees salvaged from areas being cleared for mineral sand mining on North Stradbroke Island, Queensland were transplanted (approximately 25% of the trunk of the tree was buried) using excavators (Brennan *et al.* 2005). These stags were used as perches by a range of birds, including raptors and owls. Termites and weathering are the greatest limitation to the longevity of transplanted stags. This may be slowed by treatment of the wood (e.g. arsenic treatment or tar) or by establishing a concrete base free of vegetation.

Rock outcrops were constructed from natural rocks as part of an experiment to determine whether this technique would be effective for larger scale habitat restoration for the threatened broad-headed snake (*Hoplocephalus bungaroides*) (Goldingay and Newell 2000). Most of the constructed rock outcrops were colonised by velvet geckos (the primary prey item of the broad-headed snake) and occupation was at a similar rate to comparable natural rock outcrops. The constructed rock outcrop also successfully attracted a broad-headed snake. It is important to note, nonetheless, that if the purpose is to provide habitat for a specific species, each rock will

11

Ex situ measures for conservation of flora and fauna

Ex situ measures for conserving wildlife involve taking individuals of a species out of their natural environment and managing them in an artificial environment, often so that they can be returned to the natural environment at a later time (Figure 11.1). Flora are conserved in herbariums, botanic gardens and seed banks while fauna are conserved in wildlife sanctuaries and zoos. There is scope for developers to play an important role in conservation, either by contributing financially to conservation programs (e.g. *ex situ* breeding programs for native species) or by undertaking collaborative conservation programs.

Seed preservation and translocation conducted to reduce impacts of development on wildlife is discussed in this chapter. Translocation involves moving living organisms from one area, with free release into a different area, either inside or outside the species range (International Union for Conservation of Nature 1987). Seed preservation is often a precursory measure to translocation of flora species. However, the two measures are very different.

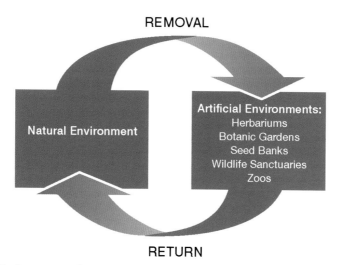

Figure 11.1. *Ex situ* measures for conserving wildlife involve taking individuals of a species out of their natural environment and managing them in an artificial environment before returning them back to nature (Diagram: J Toich)

Ex situ measures for conserving wildlife are often viewed as a last resort, because they are tagged with the perception they are expensive and provide little certainty over the conservation outcome. Nonetheless, advances in seed preservation techniques, plant propagation practices and animal breeding programs are continuing to improve *ex situ* measures. There is little doubt that maintaining viable populations of species in functioning ecosystems is much more desirable. However, *ex situ* measures for conserving wildlife do have their place – they have played a vital role in the conservation of many species, not only for preserving genetic material, but also to assist in the recovery of species in the wild (Hoyle *et al.* 1995; Cochrane 2004; Crawford and Leonie 2009).

Any *ex situ* measures for conserving wildlife must be designed with realistic objectives tailored to suit the species and situation. In many cases, *ex situ* measures for conserving wildlife should not be solely relied upon to reduce impacts, because of the uncertainty in the outcome. Notwithstanding, they can complement a range of other measures described in this book (e.g. seed preservation can complement efforts to increase the species habitat in its natural environment).

The *ex situ* measures for conserving wildlife described in this chapter (e.g. collection of native seed or translocating individuals) requires appropriate licences from regulatory authorities. The legal obligations vary from one jurisdiction to another.

Seed preservation

Overview	
What:	Preservation of seed in a recognised seed bank.
Where:	Particularly useful if the viability of a plant population is threatened by a development.
Pros:	Flora species and genetic diversity can be saved, if successful.
Cons:	Not all plant species are suitable candidates. Some collected seed may lose its ability to germinate over the long term.
Cost:	Low
Companion measures:	Plant salvage (Chapter 6); natural habitat linkages (Chapter 8); revegetation and restoration of ecosystems (Chapter 10); translocation (this chapter).
Alternative measures:	Avoid or minimise disturbance to the subject species (Chapter 4); plant salvage (Chapter 6); translocation (this chapter); conserve existing *in situ* populations by offset (Chapter 12).

Description and potential applications

Plant seed can be preserved in a seed bank for later germination. Seed banks were first created to preserve food crop species between cropping seasons. When talking about seed banks with members of the public, many people think of the Svalbard Global Seed Vault. This seed bank seems almost imaginary because it comprises large rooms built deep into the side of a mountain near the North Pole (Fowler 2008; Charles 2006). More typical seed banks in Australia comprise of a temperature controlled cool room with cabinets for seed storage, as shown in Figure 11.2. Seeds are kept in a state of 'suspended animation' in various seed banks around Australia, where they can remain viable for hundreds of years, if not longer.

Plants are a very diverse group of organisms, with an estimated 19 530 native plant species in Australia alone (Chapman 2009). Many plants and their inherent genetic diversity are

Figure 11.2.   Seeds stored in heat-sealed bags in the walk-in-freezer of the New South Wales Seed Bank, Mount Annan Botanic Gardens (Photo: Botanic Gardens Trust, Sydney)

threatened by ongoing clearance in natural areas. This is an important reason to conserve genetic material in seed banks. Seed conservation is particularly important when the viability of a flora population is at risk of extinction due to development. A seed bank can be used as an 'insurance service' for other conservation techniques, such as translocation (see next section), so that, if they turn out to be unsuccessful, the seeds still exist (Van Slageren 2003).

Seed collection commonly takes place at development sites before vegetation clearance to preserve diversity that may otherwise be lost, but also to take advantage of the seed resource for other projects so that demands on other bushland sources are reduced. It is vital that the method used to collect seed is sound, so that long-term storage will be possible. The viability of seed depends on factors such as seed maturity and seed handling methods (Gunn 2001). Seeds should be collected at the optimum times and this will influence when clearance activities can commence.

Standard procedures for collecting seed are detailed in a number of sources (e.g. Gunn 2001; Rao *et al.* 2006; Carr *et al.* 2008; Cochrane *et al.* 2009; Offord and Meager 2009). If seeds are to be collected to be banked, it is best to contact the chosen seed bank for specific collection requirements. The *FloraBank Network* (Greening Australia 2011) provides a directory of some seed banks and seed suppliers around Australia. The Seeding Victoria (2011) website provides contact details for a range of seed banks across Victoria. Otherwise, botanical gardens in each state are likely to assist. Relevant licences are required for the collection of native seed in Australia.

Offord and Meager (2009) provide the current national standard on conservation of Australian plants using seed banking. Seed banking for plant conservation gained momentum over the last 10 years as many botanical gardens and scientific institutions around Australia became partners in the Millennium Seed Bank Project and began receiving funding for seed preservation and research (Royal Botanical Gardens, Kew 2011) (Box 11.1).

Box 11.1.   Millennium Seed Bank Project – changing seed preservation in the 21st century

The Millennium Seed Bank Project was launched in 1996. The project involved the construction of the largest seed bank in the world at Kew, United Kingdom (Van Slageren 2003; Royal Botanical Gardens, Kew 2011). The project has kick-started large-scale seed preservation and research efforts all over the world, including Australia (e.g. University of Queensland's 'Seeds for Life'). The Millennium Seed Bank Project was reportedly described by Sir David Attenborough, the Project's patron, as 'perhaps the most important conservation initiative ever' (New South Wales Botanic Gardens Trust 2011).

A key objective of the Project was to safeguard 24 000 species of dryland (arid zone) plants against extinction, which equates to 10% of the world's flora (Van Slageren 2003). This objective was achieved in October 2009 and seed preservation efforts continue towards a new target: 25% of the worlds' flora by 2020 (Royal Botanical Gardens, Kew 2011). The conservation efforts focus on arid zone species for various reasons. Many arid zone species are being lost to desertification, making them among the most threatened environments on Earth. Seed from these plants lends well to seed preservation (Van Slageren 2003).

Effectiveness

Seed banks are suitable for plant species that reproduce sexually and produce seeds that can be stored at sub-zero temperatures (known as 'orthodox' seeds). Following established procedures, such as those described in Offord and Meager (2009), will maximise successful preservation of plant genetic diversity.

Not all plants are able to be preserved via seed, because some species do not produce seeds (e.g. mosses and ferns) or the seeds that are produced are recalcitrant: that is, they do not survive the drying and freezing process (Gunn 2001).

Alternative *ex situ* measures for plant conservation include the collection of living plants for planting in botanical gardens (also called field gene banks) and *in vitro* storage. *In vitro* storage is the storage of species in a laboratory, and includes propagation in test tubes and cryostorage (storage of tissue in liquid nitrogen). The high labour intensity and costs of propagation in test tubes limits its widespread use as a technique to conserve species impacted by development. Although cryostorage techniques are considerably less labour intensive and costly than propagation in test tubes, procedures for the use of cryostorage are still being trialled (see Kaczmarczyk *et al.* 2011). Cryostorage techniques have been successfully trialled on bryophytes in the United Kingdom (Rowntree *et al.* 2011). Other *ex situ* conservation measures used overseas include preservation of pollen in pollen banks and preservation of plant DNA in DNA banks (Laliberté 1997); however, these techniques are not widely used in Australia.

If a development relies on seed preservation to demonstrate that genetic diversity will not be lost, at the very least, germination testing to confirm that the seed is viable should be undertaken before plants are cleared. Generally, an initial germination test is conducted before adding seeds into a seed store. Even after germination testing, genetic changes and loss of seed viability are possible during long-term storage (Bonner 1990), so this measure is far from fail proof.

Impacts on other aspects of the environment

Because seed preservation is an *ex situ* activity, the potential impact on the environment relates largely to the collection of seed. Generally, seed collection is a relatively passive activity. However, the viability of the *in situ* plant population can become threatened if seeds are over-collected from natural areas. Collecting seeds in large quantities can also decrease food resources for animals.

Translocation

Overview	
What:	Moving living organisms from one area, with free release into a different area either inside or outside the species' range
Where:	Particularly useful where an area is to be cleared and there is no nearby alternative for relocating animals or when conservation significant plants are to be moved.
Pros:	Potential to avoid the loss of a species or genetic diversity.
Cons:	A complex process requiring a high level of management, with little certainty over the outcome in the long term.
Cost:	High
Companion measures:	Measures to provide additional habitat for flora and fauna impacted by development (Chapter 10).
Alternative measures:	Impact avoidance (Chapter 4); seed preservation (this chapter); conserve existing *in situ* populations by an environmental offset (Chapter 12).

Description and potential applications

Translocation is intended to aid the persistence of an *in situ* population of a particular species. Translocation involves either introducing an organism outside its historically known native range, re-introducing an organism in an area it formerly occurred or increasing the number of organisms in a particular area by restocking (International Union for Conservation of Nature 1987). The International Union for Conservation of Nature (1987) *Translocation of Living Organisms International Union for Conservation of Nature Position Statement* has been adopted as the basis for the main translocation guidelines in Australia (e.g. Vallee *et al.* 2004; New South Wales National Parks and Wildlife Service 2001b).

Translocation is a powerful tool used to conserve species, although, if it is misused, it can cause more harm than good. For this reason, there are set guidelines and regulations for trans-location. The *Guidelines for Translocation of Threatened Plants in Australia* (Vallee *et al.* 2004), prepared by the Australian Network for Plant Conservation, are the current standard.

Translocation is considered when a development is unable to avoid or reduce impacts, but will proceed anyway based on social and economic aspects in accordance with ecologically sustainable development. Alternative management options have also been exhausted. Alternatives include avoiding the impact altogether (Chapter 4) and conserving existing *in situ* populations within a conservation area (Chapter 12). Often, undertaking further surveys to map the distribution of seemingly rare species can reveal that the species is not as rare as previously thought, negating the need for translocation.

The re-introduction of an organism in an area it formerly occurred (e.g. a rehabilitation area) or increasing the number of organisms in a particular area by restocking are forms of translocation likely to be attempted to reduce the impact of development on a species. Translocation that involves introducing species outside their historically known native range would be rarely, if at all, relevant to development proposals, but could be considered as an adaption response to climate change.

Translocation programs need to be designed with realistic objectives, considering that there can be many limitations (e.g. lack of knowledge, important habitat components that are missing and threatening processes operating that present a risk to translocated species). The overarching aim is usually to establish or maintain a self-sustaining population of the target species that requires minimal, or no, human intervention. Translocation should only be considered if it can be demonstrated that there will be no irreparable harm to the species as a whole (Vallee *et al.* 2004) or other species.

There is scope for the translocation of plants using a number of methods (e.g. seeds, cuttings, grafting, micro-propagation, and transfer of soil containing seeds or mature plants). Translocation may also be used to re-introduce fauna species into landscapes subject to habitat deterioration, but where revegetation and restoration efforts have reinstated habitat. For example, Arid Recovery – an ecosystem restoration project in South Australia partly funded by developers as company initiatives – has included re-introductions of the greater stick-nest rat (*Leporillus conditor*), burrowing bettong (*Bettongia lesueur*), greater bilby (*Macrotis lagotis*), western barred bandicoot (*Perameles bougainville*), numbat (*Myrmecobius fasciatus*) and woma python (*Aspidites ramsayi*) (Arid Recovery 2011). In general, there is scope for more sophisticated reconstruction of ecosystems using translocation of fauna. This is particularly true for animals that have large influences over ecosystem processes (e.g. bilbies and bettongs) (James and Eldridge 2007).

Effectiveness

Short (2009) recently undertook a comprehensive review of 380 documented translocations of 102 vertebrate species within Australia. This followed a similar review completed earlier by Fischer and Lindenmayer (2000). Both reviews acknowledge that the outcomes of translocations are often not published. Short (2009), for example, noted that there was no reported outcome for 40% (152) of translocations. Of those reported, Short (2009) found that only 54% were deemed successful. Given the high chances of failure, it should not be automatically assumed that translocation programs will decrease the level of impact on *in situ* wildlife.

Some well-known translocation programs undertaken to conserve Australian species include: the northern hairy-nosed wombat (*Lasiorhinus krefftii*) (Box 11.2); bridled nailtail wallaby (*Onychogalea fraenata*) (Rout *et al* 2009); northern quoll (*Dasyurus hallucatus*) (Rankmore *et al.* 2009); and the Nielsen Park she-oak (*Allocasuarina portuensis*) (Vallee *et al.* 2004; Offord and Meager 2009).

When proposing translocation, it is worth considering why previous attempts at translocation have been successful or not, especially when proposing the translocation of a closely related species. Planning for the translocation of the northern hairy-nosed wombat (*Lasiorhinus krefftii*), for example, was streamlined by considering the reasons behind the successful translocation of the closely related southern hairy-nosed wombat (*Lasiorhinus latifrons*) (Box 11.2). Unfortunately, unsuccessful translocation attempts are less likely to be reported and published: an issue that can lead to the same mistakes being made again and again (Fischer and Lindenmayer 2000).

Translocation programs need to be planned adequately, appropriately implemented and sufficiently resourced. The effectiveness of a translocation program is improved by an understanding of:

- the species distribution and population size
- the population genetics of the species
- the species' biology (e.g. reproduction), ecological requirements and threats to its persistence
- how the translocated species may impact its new environment
- suitable sites for translocated individuals, such as land under a conservation tenure such as an offset area (Chapter 12).

Translocation can be undertaken in collaboration with institutions such as zoos and botanic gardens (Box 11.2). These institutions can sometimes assist by breeding animals in captivity or cultivating plants as stock for the translocation program.

Introducing individuals into the natural environment can prove challenging. For example, a breeding program was undertaken for the green and golden bell frog (*Litoria aurea*) at Taronga Zoo and, despite approximately 20 000 tadpoles and metamorph frogs being bred for release, all re-introductions failed (McFadden *et al.* 2008). Trials could be undertaken as a precursor to translocation to improve the effectiveness of a program. Similarly, monitoring and adaptive management will give a translocation program its best chance of success.

Impacts on other aspects of the environment

A number of potential problems can result from translocation, as shown in Figure 11.3. The potential for each of these problems to arise needs to be carefully considered for each translocation proposal. Although introducing an organism outside its historically known range has been used as a conservation tool (e.g. conserving threatened and rare mammal species by translocating them to islands in Western Australia; Abbott 2000), in some circumstances, it can have unpredictable and adverse impacts on the receiving environment (Figure 11.3).

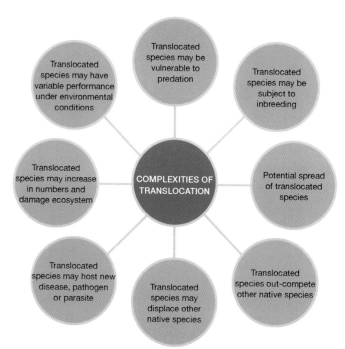

Figure 11.3.   The complexities of translocation (Diagram: J Toich)

Box 11.2.   Developers play a vital role in the conservation of the northern hairy-nosed wombat

Xstrata is a global mining company that operates mines in Queensland, New South Wales and the Northern Territory. In 2008, Xstrata pledged $3 million over 3 years towards the re-introduction program for the northern hairy-nosed wombat (*Lasiorhinus krefftii*), managed by the Queensland government (Xstrata 2008). This species is critically endangered and, prior to the re-introduction program, was restricted to a single population in an area of only 300 hectares at Epping Forest National Park near Clermont in central Queensland (Hoyle *et al.* 1995).

The re-introduction program will involve translocation of up to 24 northern hairy-nosed wombats to establish a second colony (Queensland Department of Environment and Resource Management 2011) (Figure 11.4). The success of the re-introduction program will not be known for many years.

Corporate sponsorship made the northern hairy-nosed wombat translocation program possible. The investment was not made to mitigate or offset impacts of a particular development, but rather as a company initiative. Xstrata is likely to ultimately benefit from the investment in terms of enhanced public perception of the company. The partnership between the mining company and the government demonstrates that developers can be significant contributors to conservation efforts.

Figure 11.4.   Northern hairy-nosed wombat being released into the Richard Underwood Nature Refuge (Photo: Queensland Department of Environment and Resource Management)

12

Environmental offsets

Overview	
What:	An action taken to compensate for the impacts of a development.
Where:	Useful where a proposed development will result in residual impacts that cannot be avoided or mitigated.
Pros:	Positive conservation actions in addition to avoiding and mitigating impacts.
Cons:	An environmental offset may not be able to offset the residual impact.
Cost:	High.
Companion measures:	Weed control and prevention (Chapter 9); natural habitat linkages (Chapter 8); measures to provide additional habitat for flora and fauna impacted by development (Chapter 10).
Alternative measures:	Avoiding impacts on flora and fauna (Chapter 4).

Description and potential applications

After exhausting all options to avoid and mitigate impacts on wildlife, there is usually a residual impact. Since the mid 1990s in Australia, developers have aimed to address this residual impact by providing an environmental offset, which is an action taken to compensate for the impacts of a development (Department of Environment and Water Resources 2007). Offsets for impacts on wildlife are typically the provision of parcels of land set aside for conservation purposes, although the provision of funding (e.g. for research or community wildlife projects) is sometimes granted to compensate for residual impacts. Conservation actions undertaken beyond those required to address the residual impacts are referred to as a 'company initiatives' or 'contributions to the environment' (Box 11.2).

Residual impacts vary between developments, but the loss of habitat is often the primarily residual impact. The impact on local fauna populations is another common impact requiring targeted measures to ensure their persistence in the local area. It has been argued that, to avoid loss of biodiversity, any developments that will result in the loss of natural areas should invest in restoring or enhancing other natural areas (see Lowe 2009). Although landscape-scale management arrangements may provide better conservation outcomes (Fischer *et al.* 2005;

Underwood 2011) (Box 12.1), restoring or enhancing natural areas on parcels of land set aside for conservation purposes provides the foundation for many offsets. Monetary contributions for research can sometimes lead to an acceptable conservation outcomes for species that have largely unknown habitat requirements or where land tenure is restrictive.

Government policy regarding biodiversity offsets in Australia is rapidly changing. The Australian Government has an offset policy that specifically applies to matters of national environmental significance under the Commonwealth *Environmental Protection and Biodiversity Conservation Act 1999*. Several states (Queensland, Victoria, South Australia, New South Wales and Western Australia) have developed separate offset policies, each with their own method for calculating offsets and predicting the conservation outcome.

Assessing the suitability of an offset to mitigate a residual impact requires an understanding of ecological principles and an intimate knowledge of the environment. One of the greatest challenges for government policy on offsets is providing flexibility that enables the best conservation outcome for a particular development and receiving environment (Box 12.1). Landholders and governments tend to want to simplify the transaction of land disturbed to the area of land set aside for conservation, and to this end, describe a land offset in terms of a ratio. This is a simplistic way of viewing an offset that can lead to sub-optimal conservation outcomes owing to the variety of other important factors that are overlooked when applying a ratio.

Governments have also attempted to develop mathematical programs to calculate what is required to redress residual impacts on wildlife, based on habitat measurements and presence of threatened species. Unfortunately, these types of programs, if used alone, are inherently simplistic (in relation to the complexity of ecosystems). Consequently, sub-optimal conservation outcomes may be possible.

There is a need for more adequate compliance of land offsets in Australia (Gibbons and Lindenmayer 2007). Unlike public conservation reserves, land offsets in Australia are not recorded in the Collaborative Australian Protected Areas Database (Department of Sustainability, Environment, Water, Population and Communities 2011c), using the International Union for Conservation of Nature classification system of protected areas (International Union for Conservation of Nature 2009). This is a potential issue, because the rate of land being set aside for offsets is increasing exponentially and the federal conservation databases are yet to catch up with the changes, leading to poor overall strategic management of protected areas in Australia.

Effectiveness

The effectiveness of an environmental offset is measured against the residual impact from a development. Whether the offset is land to be conserved or a monetary contribution to research or a wildlife conservation project, does it compensate for the residual impacts on wildlife? In other words, on a scale balance, would the conservation initiatives proposed outweigh the adverse impacts from the development?

It is not always possible to undertake actions to sufficiently offset residual impacts: for example, if clearing poses an immediate risk to the viability of a species (Gibbons and Lindenmayer 2007), or if the habitat proposed to be cleared is unique and/or irreplaceable. In these cases, developments sometimes still proceed based on the social and economical aspects of ecologically sustainable development. Sometimes the development may be detrimental to one species, but provides an offset that is crucial to the conservation of a range of other species. It can also be difficult to adequately offset particular residual impacts, such as fragmentation of habitat (Villarroya and Puig 2010).

It is widely acknowledged that preserving natural vegetation and habitat is a high conservation priority. Protection from current and future threats is essential to avoid losing high-quality

Box 12.1. Agricultural landscapes as potential conservation or offset areas?

Professor David Goldney
Principal Consulting Ecologist, Cenwest Environmental Services
Adjunct Professor, Charles Sturt University
Adjunct Professor, University of Sydney

Most of Australia's mines are operating within agricultural landscapes and consequently many mining companies become owners of large areas of farming and grazing land surrounding their core enterprise. These lands are purchased for a number of reasons, including for future expansion, as buffer zones to ameliorate potential impacts, such as noise and dust, or because they contain areas that can potentially be used as part of offset arrangements. Quite often, these acquired agricultural lands are leased back to individuals from the local farming community. Post-mining, some of these lands are likely to be sold at the prevailing market prices.

In Australia's oldest inland agricultural land in the Central West of New South Wales, recent research has demonstrated the following:

- Land degradation and biodiversity losses continue along an exponential pathway, but repair and restoration funding that is provided mainly by governments is on an *ad hoc* basis and, at best, follows a linear trajectory. This results in the gap between what is required and what is provided continuing to widen.
- Ecosystem and landscape malfunction are widespread, and their repair is usually not considered in most restoration strategies.
- Thirteen per cent of vertebrate species are listed as Vulnerable or Endangered. However, 57% of the remaining species have been assessed as regionally endangered, but this is not reflected in legislation.
- It is very likely that the outcomes described for the Central West of NSW will be repeated across agricultural Australia.

Because landholders in Australia manage around 75% of our land mass, the majority as grazing enterprises, Australian farmers and graziers are pivotal in repairing degraded landscapes and restoring and maintaining biodiversity (species diversity, appropriate habitats and ecosystem functions and cycles). However, most landholders cannot find the needed resources to implement appropriate restoration and biodiversity conservation measures in a timely manner and, furthermore, they often lack the needed skills. In contrast, mining companies usually have the needed resources to upgrade their land management practices in their agricultural holdings.

Optimising both biodiversity and production outcomes in mining buffer lands through innovative adoption of holistic farming systems (e.g. the use of cell grazing to provide 100% ground cover and/or to drive vegetation succession to favour native grasses over exotic grasses) would contribute significantly towards achieving sustainable farming systems and concurrently increase biodiversity values of the farmland (species diversity, ecosystem services and ecosystem processes). Such intervention strategies could also be accompanied by the repair of key ecosystem cycles and the conversion of degrading landscapes into aggrading landscapes. Example of this would be the re-establishment of reed beds in degraded first-order streams and maintaining near 100% ground cover across grazing land to ensure that optimal rainfall is retained in the landscape and concurrently sediments (potential resources) are trapped within the farm-scape, rather than lost through erosive processes.

Such restoration programs need to be based on agriculture as an applied ecology rather than agriculture as an applied technology. Two of the most important initiatives that are likely to follow the adoption of an ecological farming world view would be the creation of appropriate conditions needed to increase the numbers and diversity of useful soil organisms and to raise soil carbon levels. Both initiatives are necessary precursors for the development of healthy soils that underwrite sustainable farming and biodiversity conservation. Adopting holistic farming methods within degraded mining buffer lands is likely to provide a significant benefit for the environment. Similarly, proposed offset land with a balance of degraded agricultural land and areas of varying quality of remnant bushland and/or creek-lines that can respond positively to restoration initiatives has the potential to provide a significant long-term net gain in biodiversity values. In contrast, the purchase of near pristine conservation land would have little scope for improving inherent conservation values.

habitat, particularly where the habitat is already degrading. However, some offsets that involve locking up natural vegetation in pristine condition result in little conservation gain (Gibbons and Lindenmayer 2007). This is especially true if the area is unlikely to be cleared anyway because, for example, the land is too steep to be developed.

Improving the condition of degraded vegetation, by managing threatening processes, for example, can achieve a greater conservation gain, because additional habitat is created. The ability of a degraded community to regenerate naturally cannot be generalised. An assessment of the resilience of the land is required. For example, a 100-year-old cropped paddock will have very low resilience if crops are removed because there will be poor storages of native seeds in the seed bank. On the other hand, a native pasture grazed intermittently would have moderately high resilience because the seed bank should contain more native seeds. Some communities have a remarkable ability to regenerate in response to seasonal rainfall (e.g. semi-arid chenopod shrublands); however, others may regenerate into a dense monoculture of saplings.

Revegetation of cleared areas remains an important tool for conservation biology as described in Chapter 10. Incorporating revegetation into an offset proposal can accelerate habitat enhancement and restore landscape function in cleared low resilient land. Bennett *et al.* 2000 provide a simple overview of the principles for using revegetation for wildlife conservation, including reducing edge effects, positioning vegetation to increase wildlife movement across the landscape and integrated land use. An offset generally should not comprise solely of revegetating cleared land, because there is always a time lag in establishing vegetation (e.g. Maron *et al.* 2010). Also, revegetation isn't likely to provide the full biodiversity values of complex ecosystems (Hobbs *et al.* 2011). On the other hand, an offset that combines revegetation of cleared low resilient land with existing vegetation can result in high conservation gain (see Lindenmayer 2011), especially at the landscape level when natural habitat linkages are repaired. Revegetation often comes down to a question of cost–benefit.

The location of an offset area is an important consideration when determining whether it mitigates the residual impact from a development. Is the offset area suitability located to benefit the local population that is in need of additional habitat or will the offset benefit a different population that may not be at threat? It is true that, if the long-term viability of a local population is questionable, it may be best to focus conservation efforts on more certain conservation outcomes to benefit the overall species. Land conserved for the purposes of off-setting development should also be strategically located to complement an existing conservation reserve system.

13

Monitoring

How can monitoring be used?

The term 'monitoring' is widely used in the context of reducing impacts of development on wildlife. It is, however, not a measure to reduce impacts, but rather a tool that can inform management. In fact, monitoring that cannot inform management (e.g. due to poor design) might as well not be undertaken (see Field *et al.* 2007).

Monitoring can be undertaken before implementing measures as a way of determining whether they are required (i.e. if wildlife begin to be adversely impacted by a development). Monitoring may reveal that a species is continuing to live near a development site and is not being adversely impacted by the development. In this case, monitoring has shown that other more-elaborate measures to reduce impacts are not required.

The best way to determine whether a particular measure effectively reduces the impacts of development on wildlife for a given situation is to monitor its success. The results of monitoring, if effectively executed, will reveal whether the selected measure needs to be adjusted in any way or whether an alternative measure should be used instead. Importantly, monitoring can be used to confirm whether a measure used to reduce impacts on wildlife continues to be successful over time. In this case, it is important to include baseline data (i.e. data collected before the measure is implemented) in order to determine if the measure is actually effective.

Ecological monitoring approaches

The level of monitoring required will vary according to each situation and may range from the simple (e.g. directing a camera on an artificial nesting platform to monitor its use) to the complex (e.g. monitoring to determine whether the provision of a large number of artificial hollows has resulted in an increase in population of a threatened species). There are many different techniques that have been used to monitor the effectiveness of the measures and it is beyond the scope of this book to outline all monitoring methods. However, it is important that monitoring methodology is systematic and repeatable (Denny 2010).

A lack of well-defined questions at the beginning of a monitoring program can lead to failure of the monitoring program (e.g. Lindenmayer and Likens 2010a; Lindenmayer and Likens 2010b). If a monitoring program is aimed at confirming whether a measure is successful, it needs to be tailored to the objective of the measure. For example, is the objective of a chosen measure to avoid any impacts on a species or is a particular type or level of impact acceptable?

In the case of a nesting platform, the most obvious question would be 'is the nesting platform being used by the target species – yes or no?' Naturally, more complex objectives would require more than one question. If the monitoring approach is complex, a statistically robust design should be formulated before the monitoring approach is finalised, to enable data to be analysed.

Monitoring should not itself have an impact on the wildlife being monitored (e.g. monitoring undertaken too frequently can cause problems). In some cases, the monitoring approach may need to be changed. Adaptive monitoring has recently been proposed as an approach for effective ecological monitoring, involving changes to monitoring parameters over time (Lindenmayer and Likens 2010a). Care is needed when attempting to apply an adaptive monitoring approach, however, because of any data incompatibilities that may arise.

Regular inspection to see whether the measure is in need of maintenance is termed here as 'maintenance monitoring'. For a number of measures in this book, we have noted that frequent maintenance monitoring is required so that issues can be responded to quickly. The effectiveness of a fauna exclusion fence, for example, is reliant on regular inspections for breaches and their quick repair (Chapter 7). Similarly, artificial tree hollows will need to be monitored over an extended period of time; otherwise, the measure could fail within a few months if action is not taken to rectify any problems (e.g. the next box falls from the tree or is overrun with ants).

Monitoring results

The results of monitoring should ideally be shared so that measures to reduce impacts of development on wildlife can be improved and applied with greater confidence for future projects. Rigorous monitoring may be suitable for publishing in scientific journals with a world-wide audience.

14

Adaptive management

Rigorous monitoring and evaluation and, in less complex circumstances, practical experience can reveal that a measure is not working effectively and needs to be adjusted or replaced with an alternative measure. Progressively improving management through learning is the process of 'adaptive management'. The purpose of adaptive management is to gain knowledge and use it to modify practices to achieve management goals (Lindenmayer and Burgman 2005).

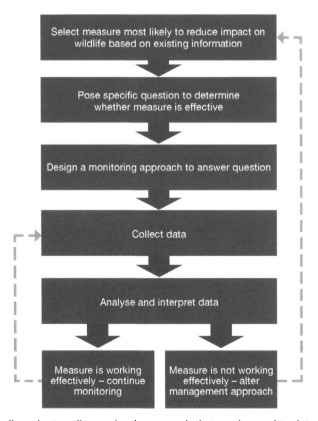

Figure 14.1. This flow chart outlines a simple approach that can be used to determine whether a measure is effective and what to do if it is working or if it is not (Diagram: J Toich)

Evaluating management options is a key component of adaptive management. If a measure to reduce impacts on wildlife is not being used as intended by the target species (e.g. a nesting platform is not being used by an osprey), a decision needs to be made whether to continue to wait for the species to use the measure or to change the approach (e.g. modify the platform or relocate it). Alternative management options could be considered from the outset of monitoring or implementing the measure; however, entirely new options may need to be considered if the alternatives considered do not tackle the issue (e.g. if non-target, but valued, fauna are using the nesting platform, a suitable adaptive management approach may be to construct another nesting platform). A sophisticated adaptive management process enables all available management options to be assessed and the most appropriate option to be selected. This process is much more informative than simple trial and error, and may involve specific studies to evaluate the effectiveness of management options (e.g. Mackenzie and Keith 2009).

Collecting data that are not used when making management decisions can be wasteful. For example, in one case, changes in vegetation were monitored over a 30-year period, but, despite the data supporting the need for ecological burning, no burning was attempted during the monitoring period (Benson and Picone 2009).

A simple adaptive management approach that can be used to determine whether a measure is effective and what to do if it is working, or if it is not, is outlined in Figure 14.1.

Concluding remarks

The measures investigated in this book have all been tried somewhere in Australia or overseas at one time or another. We have shown that measures can be applied to tackle similar impacts on wildlife across different types of development and across different states and territories in Australia. Broadly speaking, however, the effectiveness of all of the measures will be context-dependent. A measure may be effective for a particular species in a particular situation, but not for the same species in a different situation.

We have described how many of the measures in this book can be classed as 'effective', because they reduce impacts at the level of the individual animal or plant. A fauna underpass, for example, has been shown to be used by animals to cross under a busy motorway. Conserving individual plants and animals is important, especially from an ethical point of view. Yet, for the majority of the measures, it is still not known whether they are effective at maintaining plant and animal populations, particularly over the long term. This type of information is crucial because maintaining populations goes a long way towards preventing species from becoming extinct.

Just as development has accelerated in Australia over recent years, research focusing on the effectiveness of measures to reduce the impact of development on wildlife needs to increase proportionally. Developers can, and often do, play a large role in conserving populations by funding research, undertaking conservation projects or by becoming involved in a collaborative way. Developers who contribute to conservation of wildlife are bound to be rewarded in the form of enhanced public perception of their company. By targeting this research around development scenarios, it can also help avoid unnecessary costs of measures imposed on them and enable more certainty about how a development can operate effectively and still avoid, minimise and offset impacts on wildlife.

As existing measures to reduce impacts on wildlife are tested more rigorously, and new measures are developed, more will become apparent about their effectiveness. It is important that this knowledge is shared and published in a form that can be readily accessed by peers. Ideally, this would be via scientific journals, but not all knowledge will be suitable for publication in journals, and other avenues may be more appropriate. The framework for *gathering* information is already in place. Through government assessment and regulation of developments, measures are being implemented, data is being collected and reports are being written.

We stated earlier that we hope that this book will energise the topic and inspire others to come up with their own innovative ideas. These may be variations to the existing measures or entirely new measures. New measures should be implemented experimentally and present a low risk to wildlife if they fail.

We believe this to be the first book entirely dedicated to reducing a wide range of impacts of development on wildlife and, in this respect, this book is just the starting point. We hope that readers will share their experiences with us so that we are able to make an even better revision of this book in the years to come (contact us at gleesonecology@bigpond.com). It is by sharing knowledge, and then implementing improved measures, that impacts of development on wildlife will be reduced.

References

Aaronson RJ (2007) Leadership and action: a winning combination. *International Civil Aviation Organisation Journal* **5**, 36.

Abbott I (2000) Improving the conservation of threatened and rare mammal species through translocation to islands: case study Western Australia. *Biological Conservation* **93**, 195–201.

ABC (Australian Broadcasting Corporation) News (2004) *Mass Animal Deaths Spark Crackdown*. Australian Broadcasting Corporation News, <http://www.abc.net.au/news/stories/2004/03/23/1071910.htm>.

Abel N, Baxter J, Campbell A, Cleugh H, Fargher J, Lambeck R et al. (1997) Design Principles for Farm Forestry – A Guide to Assist Farmers to Decide Where to Place Trees and Farm Plantations on Farms. Rural Industries Research and Development Corporation, Canberra.

ABS Energy Research (2008) *The T&D Report, Seventh Edition*. ABS Energy Research, London.

Adkins MF (2006) A burning issue: using fire to accelerate tree hollow formation in Eucalypt species. *Australian Forestry* **69**, 107–113.

Al-Ghamdi AS and AlGadhi SA (2004) Warning signs as countermeasures to camel-vehicle collisions in Saudi Arabia. *Accident Analysis and Prevention* **36**, 749–760.

Allen GT and Ramirez P (1990) A review of bird deaths on barbed wire. *The Wilson Bulletin* **102**, 553–558.

Alonso JC, Alonso JA and Munoz-Pulido R (1994) Mitigation of bird collisions with transmission lines through groundwire marking. *Biological Conservation* **67**, 129–134.

Anderson VJ and Hodgkinson KC (1997) Grass-mediated capture of resource flows and the maintenance of banded Mulga in a semi-arid woodland. *Australian Journal of Botany* **45**, 331–342.

Andersson M, Wallander J and Isaksson D (2009) Predator perches: a visual search perspective. *Functional Ecology* **23**, 373–379.

Andrews KM, Gibbons JW and Jochimsen DM (2008) Ecological effects of roads on amphibians and reptiles: a literature review. *Herpetology Conservation* **3**, 121–143.

Anstis M (2002) *Tadpoles of South-eastern Australia – A Guide with Keys*. New Holland Publishers (Australia), Sydney.

Ardia DR, Perez JH and Clotfelter ED (2006) Nest box orientation affects internal temperature and nest site selection by Tree Swallows. *Journal of Field Ornithology* **77**, 339–344.

Arid Recovery (2011) *Welcome to Arid Recovery*. Arid Recovery, South Australia, <http://www.aridrecovery.org.au>.

Arida EA and Bull CM (2008) Optimising the design of artificial refuges for the Australian skink, *Egernia stokesii*. *Applied Herpetology* **5**, 161–172.

Armstrong KN (2001) The distribution and roost habitat of the orange leaf-nosed bat *Rhinonicteris aurantius*, in the Pilbara region of Western Australia. *Wildlife Research* **28**, 95–104.

Armstrong KN and Anstee SD (2000) The ghost bat in the Pilbara: 100 years on. *Australian Mammalogy* **22**, 93–101.

Asari Y, Johnson CN, Parsons M and Larson J (2010) Gap-crossing in fragmented habitats by mahogany gliders (*Petaurus gracilis*), do they cross roads and powerline corridors? *Australian Mammalogy* **32**, 10–15.

Attiwill P and BA Wilson (2006) *Ecology: An Australian Perspective*. Oxford University Press, Melbourne.

Auld TD and Keith DA (2009) Dealing with threats: integrating science and management. *Ecological Society of Australia* **10**, 79–87.

Australian Bureau of Statistics (2008) *Population Projections, Australia, 2006 to 2101*. Australian Bureau of Statistics, <http://www.abs.gov.au/Ausstats/abs@.nsf/mf/3222.0>.

Australian Bureau of Statistics (2011a) *Australia's Demographic Statistics, December 2010*. Australian Bureau of Statistics, <http://www.abs.gov.au/ausstats/abs@.nsf/Latestproducts/3101.0Media%20Release1Dec%202010?opendocument&tabname=Summary&prodno=3101.0&issue=Dec%202010&num=&view>.

Australian Bureau of Statistics (2011b) *Australia's Environment: Issues and Trends, Jan 2010*. Australian Bureau of Statistics, <http://www.abs.gov.au/AUSSTATS/abs@.nsf/Lookup/4613.0Chapter95Jan+2010>.

Australian Department of Agriculture, Fisheries and Forestry (2008) *Rescue and Rehabilitation of Sick, Injured or Orphaned Wildlife*. Australian Department of Agriculture, Fisheries and Forestry, <http://www.daff.gov.au/animal-plant-health/welfare/nccaw/guidelines/wildlife/rescue>.

Australian Pipeline Industry Association (2009) *Code of Environmental Practice – Onshore Pipelines*. Australian Pipeline Industry Association, Canberra.

Australian Transport Safety Bureau (2003) *The Hazard Posed to Aircraft by Birds. Research Paper November 2002*. Australian Transport Safety Bureau, Canberra.

Australian Transport Safety Bureau (2008) *An Analysis of Australia Bird Strike Occurrences 2002 to 2006*. Australian Transport Safety Bureau, Canberra.

Australian Transport Safety Bureau (2010) *Australian Aviation Wildlife Strike Statistics: Bird and Animal Strikes 2002 to 2009*. Australian Transport Safety Bureau, Canberra.

Balkenhol N and Waits LP (2009) Molecular road ecology: exploring the potential of genetics for investigating transportation impacts on wildlife. *Molecular Ecology* **18**, 4151–4164.

Ball TM and Goldingay RL (2008) Can wooden poles be used to reconnect habitat for a gliding mammal? *Landscape and Urban Planning* **87**, 140–146.

Ballantyne R and Hughes K (2006) Using front-end and formative evaluation to design and test persuasive bird feeding warning signs. *Tourism Management* **27**, 235–246.

Bamberger Ranch Preserve (2011) *The Bats of Selah*. Bamberger Ranch Preserve, Texas, <http://www.bambergerranch.org/news/bats.phtml>.

Bank FG, Irwin CL, Evink GL, Gray ME, Hagood S, Kinar JR *et al.* (2002) *Wildlife Habitat Connectivity Across European Highways*. Technical Report FWHA-PL-02-011. US Department of Transportation, Washington DC.

Barlow GB and Bock K (1984) Predation on fish farms by cormorants *Phalacrocorax* spp. *Australian Wildlife Research* **11**, 559–566.

Barras SC and Seamans TW (2002) Habitat management approaches for reducing wildlife use of airfields. 4–7 March, Nevada, USA. In *Proceedings of the Twentieth Vertebrate Pest Conference*, (Eds RM Timm) University of California, Davis.

Barton JL and Davies PE (1993) Buffer strips and streamwater contamination by atrazine and pyrethroids aerially applied to *Eucalyptus nitens* plantations. *Australian Forestry* **56**, 201–210.

Bat Conservation International (2011) *Criteria for Successful Bat Houses*. Bat Conservation International, <http://www.batcon.org/>.

Baudinette RV (1994) Microclimate conditions for maternity caves of the Bent-wing Bat, *Miniopterus schereibersii*: an attempted restoration of a former maternity site. *Wildlife Research* **21**, 607–619.

Baudinette RV, Churchill SK, Christian KA, Nelson JE and Hudson PJ (2000) Energy, water balance and the roost microenvironment in three Australian cave-dwelling bats (Microchiroptera). *Journal of Comparative Physiology B* **170**, 439–446.

Bax D (2006) *Karuah Bypass Fauna Crossing Report*. Thiess Pty Ltd, Sydney.

Baxter AT and Allan JR (2006) Use of raptors to reduce scavenging bird numbers at landfill sites. *Wildlife Society Bulletin* **34**, 1162–1168.

Beatly T and Newman P (2009) Green Urbanism Down Under – Learning from Sustainable Communities in Australia. Island Press, Washington DC.

Beeh P (1994) Bumpy ride for forest cableway – tours over the treetops – ecological disaster or an economic necessity. *Geo Australasia* **16**, 24–29.

Beeton RJS, Buckley KI, Jones GJ, Morgan D, Reichelt RE and Trewin D (2006 Australian State of the Environment Committee) (2006) *Australia State of the Environment 2006*. Independent report to the Australian Government Minister for the Environment and Heritage, Department of the Environment and Heritage, Canberra.

Beger M, Grantham HS, Pressey RL, Wilson KA, Peterson EL, Dorfman D *et al.* (2010) Conservation planning for connectivity across marine, freshwater, and terrestrial realms. *Biological Conservation* **143**, 565–575.

Begon M, Townsend CR and Harper JL (2005) *Ecology: From Individuals to Ecosystems, Fourth Edition*. Wiley-Blackwell, Brisbane.

Beier P and Loe S (1992) In my experience: a checklist for evaluating impacts to wildlife movement corridors. *Wildlife Society Bulletin* **20**, 434–440.

Beier P and Noss RF (1998) Do habitat corridors provide connectivity? *Conservation Biology* **6**, 1241–1252.

Belant JL and Ickes SK (1996) Overhead wires reduce roof-nesting by ring-billed gulls and herring gulls. In *Proceedings of the 17th Vertebrate Pest Conference*. 5–7 March, Rohnert Park, California. (Eds RM Timm and AC Crabb) p. 108. University of California, Davis.

Bender H (2001) Deterrence of Kangaroos from Roadways using Ultrasonic Frequencies – Efficacy of the Shu Roo. University of Melbourne, Melbourne.

Bender H (2003) Deterrence of kangaroos from agricultural areas using ultrasonic frequencies: efficacy of a commercial device. *Wildlife Society Bulletin* **31**, 1037–1046.

Bender H (2005) Effectiveness of the eastern grey kangaroo foot thump for deterring conspecifics. *Wildlife Research* **32**, 649–655.

Bennett AF (1991) Roads, roadsides and wildlife conservation: a review. In *Nature Conservation 2: The Role of Corridors*. (Eds DA Saunders and RJ Hobbs) pp. 99–117. Surrey Beatty & Sons, Chipping Norton, UK.

Bennett AF, Kimber SL and Ryan PA (2000) Revegetation and Wildlife – A Guide to Enhancing Revegetated Habitats for Wildlife Conservation in Rural Environments. Environment Australia, Canberra.

Bennett AF, Haslem A, Cheal DC, Clark MF, Jones RN, Koehn JD *et al.* (2009) Ecological processes: a key element in strategies for nature conservation. *Ecological Management and Restoration* **10**, 192–199.

Benson D and Picone D (2009) Monitoring vegetation change over 30 years: lessons from an urban bushland reserve in Sydney. *Cunninghamia* **11**, 195–202.

Berge AJ, Delwiche MJ, Gorenzel WP and Salmon TP (2007) Sonic broadcast unit for bird control in vineyards. *Applied Engineering in Agriculture* **23**, 819–825.

Bevanger K (1994) Bird interactions with utility structures: collision and electrocution, causes and mitigating measures. *IBIS* **136**, 412–425.

Bevanger K (1998) Biological and conservation aspects of bird mortality caused by electricity power lines: a review. *Biological Conservation* **86**, 67–76.

Beyer GL and Goldingay RL (2006) The value of nest boxes in the research and management of Australian hollow-using arboreal marsupials. *Wildlife Research* **33**, 161–174.

Biedenweg TA, Parsons MH, Fleming PA and Blumstein DT (2011) Sounds scary? Lack of habituation following the presentation of novel sounds. *PLoS ONE* **6**, 1–8.

Bishop J, McKay H, Parrott D and Allan J (2003) Review of International Research Literature Regarding the Effectiveness of Auditory Bird Scaring Techniques and Potential Alternatives. Report for Department for Environment Food and Rural Affairs, <http://www.defra.gov.uk/environment/quality/noise/research/birdscaring/birdscaring.pdf>.

Bishop-Hurley GJ, Swain DL, Anderson DM, Sikka P, Crossman C and Corke P (2007) Virtual fencing applications: implementing and testing an automated cattle control system. *Computers and Electronics in Agriculture* **56**, 14–22.

Bissonette JA and Hammer M (2000) Effectiveness of earthen return ramps in reducing big game highway mortality in Utah. *UTCFWRU Report Series 2000* **1**, 1–29.

Blackwell BF, DeVault TL, Fernandez-Juricic E and Dolbeer RA (2009) Wildlife collisions with aircraft: a missing component of land-use planning for airports. *Landscape and Urban Planning* **93**, 1–9.

Blumstein DT, Anthony LL, Harcourt R and Ross G (2003) Testing a key assumption of wildlife buffer zones: is flight initiation distance a species-specific trait. *Biological Conservation* **110**, 97–100.

Boag DA and Lewin V (1980) Effectiveness of three waterfowl deterrents on natural and polluted ponds. *The Journal of Wildlife Management* **44**, 145–154.

Bomford M (1990) Ineffectiveness of a sonic device for deterring starlings. *Wildlife Society Bulletin* **18**, 151–156.

Bomford M and O'Brien PH (1990) Sonic deterrents in animal damage control: a review of device tests and effectiveness. *Wildlife Society Bulletin* **18**, 411–422.

Bomford M and Sinclair R (2002) Australian research on bird pests: impact, management and future directions. *Emu* **102**, 29–45.

Bond AR and Jones DN (2008) Temporal trends in use of fauna-friendly underpasses and overpasses. *Wildlife Research* **35**, 103–112.

Bonner FT (1990) Storage of seeds: potential and limitations for germplasm conservation. *Forest Ecology and Management* **35**, 35–43.

Booth C (2007a) *Barbed Wire Action Plan,* Queensland Conservation, <http://www.wildlifefriendlyfencing.com/action.htm>.

Booth C (2007b) Barbs and birds. *Wingspan* **17**, 9–15.

Brainwood M and Burgin S (2009) Hotspots of biodiversity or homogeneous landscapes? Farm dams as biodiversity reserves in Australia. *Biodiversity Conservation* **18**, 3043–3052.

Bramley GN and Waas JR (2001) Laboratory and field evaluation of predator odors as repellents for kiore (*Rattus exulans*) and ship rats (*Rattus rattus*). *Journal of Chemical Ecology* **27**, 1029–1047.

Brennan KEC, Nichols OG and Majer JD (2005) *Innovation Techniques for Promoting Fauna Return to Rehabilitated Sites Following Mining.* Australian Centre for Minerals Extension and Research (ACMER), Brisbane and Minerals and Energy Research Institute of Western Australia (MERIWA), Perth.

Briggs SV, Taws NM, Seddon JA and Vanzella B (2008) Condition of fenced and unfenced remnant vegetation in inland catchments in south-eastern Australia. *Australian Journal of Botany* **56**, 590–599.

Briot JL (2005) *Last Experiments with a Laser Equipment Designed for Avian Dispersal in Airport Environment.* Paper presented at the International Bird Strike Committee, <http://int-bird strike.org/Athens_Papers/IBSC27%20WPV-1.pdf>.

Brough T and Bridgeman CJ (1980) An evaluation of long grass as a bird deterrent on British airfields. *Journal of Applied Ecology* **17**, 243–253.

Brouwer D and Young R (1998) Design of farm dam wetlands. In *The Constructed Wetlands Manual.* (Ed. New South Wales Department of Land and Water Conservation), Sydney, New South Wales Department of Land and Water Conservation, Sydney.

Buchanan BW (1993) Effects of enhanced lighting on the behaviour of nocturnal frogs. *Animal Behaviour* **45**, 893–899.

Buchanan RA (2009) *Restoring Natural Areas in Australia.* Tocal College, New South Wales Department of Industry and Investment, Patterson.

Bull EL and Partridge AD (1986) Methods of killing trees for use by cavity nesters. *Wildlife Society Bulletin* **14**, 142–146.

Bunn SE (2008) Review of Environmental Impact Statement and Supplementary Materials on Proposed Traveston Crossing Dam, Mary River, South-east Queensland: II: Final Report. Prepared for the Department of the Environment, Water, Heritage and the Arts, Canberra.

Burns EL, Eldridge MDB and Houlden BA (2004) Microsatellite variation and population structure in a declining Australian Hylid *Utoria aurea. Molecular Ecology* **13**, 1745–1757.

Butler MW, Whitman BA and Duffy AM (2009) Nest box temperature and hatching success of American Kestrels varies with nest box orientation. *Wilson Journal of Ornithology* **121**, 778–782.

Cale PG (2003) The influence of social behaviour, dispersal and landscape fragmentation on population structure in a sedentary bird. *Biological Conservation* **109**, 237–248.

Carey AB and Sanderson HR (1981) Routine to accelerate tree cavity formation. *Wildlife Society Bulletin* **9**, 14–21.

Carr D, Rawlings K and Atkinson P (2008) *Native Vegetation Management Tool, Greening Australia,* <http://www.florabank.org.au/default.asp?V_DOC_ID=954>.

Chapman AD (2009) *Numbers of Living Species in Australia and the World, Second Edition.* Report for the Australian Biological Resources Study, Canberra, <http://www.environment.gov.au/biodiversity/abrs/publications/other/species-numbers/index.html>.

Chapman T (2007) Foods of the Glossy Black-cockatoo (*Calyptorhynchus lathami*). *Australian Field Ornithology* **24**, 30–36.

Charles D (2006) A 'forever' seed bank takes root in the arctic. *Science* **312**, 1730–1731.

Churchill S (2008) *Australian Bats, Second Edition.* Jacana Books, New South Wales.

Churchill SK (1991) Distribution, abundance and roost selection of the Orange Horseshoe-bat, *Rhinonycteris aurantius,* a tropical cave dweller. *Wildlife Research* **18**, 343–353.

Churchill SK (1995) Reproductive ecology of the Orange Horseshoe bat, *Rhinonycteris aurantius* (Hipposideridae: Chiroptera), a tropical cave dweller. *Wildlife Research* **22**, 687–698.

Clarke DJ and White JG (2008) Towards ecological management of Australian power line corridor vegetation. *Landscape and Urban Planning* **86**, 257–266.

Clevenger AP and Sawaya MA (2010) Piloting a non-invasive genetic sampling method for evaluating population-level benefits of wildlife crossing structures. *Ecology and Society* **15**, 7.

Clevenger AP, Chruszcz and Gunson K (2001) Drainage culverts as habitat linkages and factors affecting passage by mammals. *Journal of Applied Ecology* **38**, 1340–1349.

Cochrane A (2004) Western Australia's ex situ program for threatened species: a model integrated strategy for conservation. In *Ex situ Plant Conservation: Supporting Species Survival in the Wild.* (Eds EO Guerrant, K Havens and M Maunder) pp. 40–66. Island Press, Washington DC.

Cochrane A, Crawford AD and Offord CA (2009) Seed and vegetative material collection. In *Plant Germplasm Conservation in Australia – Strategies and Guidelines for Developing, Managing and Utilising Ex situ Collections* (Eds CA Offord and PF Meager) pp. 35–62. Australian Network for Plant Conservation Inc., Canberra.

Cockle KL and Richardson JS (2003) Do riparian buffer strips mitigation the impacts of clearcutting on small mammals. *Biological Conservation* **113**, 133–140.

Codd JR, Sanderson KJ and Branford AJ (2003) Roosting activity budget of the southern bent-wing bat (*Miniopterus schreibersii bassanii*). *Australian Journal of Zoology* **51**, 307–316.

Coffin AW (2007) From roadkill to road ecology: a review of the ecological effects of roads. *Journal of Transport Geography* **15**, 396–406.

Cole I and Lunt ID (2005) Restoring Kangaroo Grass (*Themeda triandra*) to grassland and woodland understoreys: a review of establishment requirements and restoration exercises in south-east Australia. *Ecological Management and Restoration* **6**, 28–33.

Commonwealth of Australia (2007a) Biodiversity Management – Leading Practice Sustainable Development Program for the Mining Industry. Commonwealth of Australia, Canberra.

Commonwealth of Australia (2007b) *Lighting the Way – A Local Government Guide to Energy Efficient Public Lighting on Minor Roads*. Australian Greenhouse Office in the Department of the Environment and Water Resources, Canberra.

Commonwealth of Australia (2008) Cyanide Management – Leading Practice Sustainable Development Program for the Mining Industry. Commonwealth of Australia, Canberra.

Connell JH (1978) Diversity in tropical rainforests and coral reefs. *Science* **199**, 1302–1310.

Connolly TA, Day TD and King CM (2009) Estimating the potential for reinvasion by mammalian pests through pest-exclusion fencing. *Wildlife Research* **36**, 410–421.

Conover MR (2001) *Resolving Human-Wildlife Conflicts: The Science of Wildlife Damage Management,* Lewis Publishers, Boca Raton, Florida.

Cook A, Rushton S, Allan J and Baxter A (2008) An evaluation of techniques to control problem bird species on landfill sites. *Environmental Management* **41**, 834–843.

Cook GD and Dawes-Gromadzki TZ (2005) Stable isotope signatures and landscape functioning in banded vegetation in arid-central Australia. *Landscape Ecology* **20**, 649–660.

Corlatti L, Hacklander K and Frey-roos F (2009) Ability of wildlife overpasses to provide connectivity and prevent genetic isolation. *Conservation Biology* **23**, 548–556.

Coulson GM (1982) Road-kills of macropods on a section of highway in central Victoria. *Australian Wildlife Research* **9**, 21–26.

Crawford A and Leonie M (2009) The road to recovery: the contribution of seed conservation and reintroduction to species recovery in Western Australia. *Journal of the Australian Network for Plant Conservation* **17**, 15–17.

CRC for Australian Weed Management (2004) *Introductory Weed Management Manual*. Natural Heritage Trust. CRC for Australian Weed Management, Canberra.

Croak BM, Pike DA, Webb JK and Shine R (2010) Using artificial rocks to restore nonrenewable shelter sites in human-degraded systems: colonization by fauna. *Restoration Ecology* **18**, 428–438.

Commonwealth Scientific and Industrial Research Organisation (2011) *Fencing Livestock In – Virtually.* CSIRO, Melbourne, < http://www.csiro.au/science/Virtual-Fencing-Project.html>.

Cunningham RB, Lindenmayer DB, Crane M, Michael D and MacGregor C (2007) Reptile and arboreal marsupial response to replanted vegetation in agricultural landscapes. *Ecological Applications* **17**, 609–619.

Damschen EI, Haddad NM, Orrock JL, Tewksbury JJ and Levey DJ (2006) Corridors increase plant species richness at large scales. *Science* **313**, 1284–1286.

D'Angelo GJ, D'Angelo JG, Gallagher GR, Osborn DA, Miller KV and Warren RJ (2006) Evaluation of wildlife warning reflectors for altering white-tailed deer behavior along roadways. *Wildlife Society Bulletin* **34**, 1175–1183.

Darcovich K and O'Meara J (2008) An Olympic legacy: green and golden bell frog conservation at Sydney Olympic Park 1993–2006. *Australian Zoologist* **34**, 236–248.

Date EM, Ford HA and Recher HF (1991) Frugivorous pigeons, stepping stones, and weeds in northern New South Wales. In *Nature Conservation 2: The Role of Corridors*. (Eds DA Saunders and RJ Hobbs) pp. 241–246. Surrey Beatty & Sons, Chipping Norton, UK.

Davidson I, Scammell A, O'Shannassy P, Mullins M and Learmonth S (2005) Travelling stock reserves: refuges for stock and biodiversity? *Ecological Management and Restoration* **6**, 5–15.

Day T and MacGibbon R (2007) 'Multiple-species exclusion fencing and technology for mainland sites'. USDA National Wildlife Research Center Symposia Managing Vertebrate Invasive Species. University of Nebraska – Lincoln.

Denny M (2010) Then and now – fauna monitoring within the Sydney Basin. In *A Natural History of Sydney*. (Eds D Lunney, P Hutchings and D Hochuli) pp. 90–101. Royal Zoological Society of NSW, Sydney.

Department of Environment and Water Resources (2007) *Draft Policy Statement: Use of Environmental Offsets under the Environment Protection and Biodiversity Conservation Act 1999*. Commonwealth of Australia, Canberra, <http://www.environment.gov.au/epbc/publications/pubs/draft-environmental-offsets.pdf>.

Department of the Environment and Heritage (2006) *Threat Abatement Plan Infection of Amphibians with Chytrid Fungus Resulting in Chytridiomycosis*. Department of the Environment and Heritage, Canberra.

Department of the Environment, Water, Heritage and Arts (2009a) *Significant Impact Guidelines for the Vulnerable Green and Golden Bell Frog (Litoria aurea)*. Commonwealth of Australia, Canberra, < http://www.environment.gov.au/epbc/publications/pubs/litoria-aurea-policy.pdf>.

Department of the Environment, Water, Heritage and Arts (2009b) *Significant Impact Guidelines for the Critically Endangered Spiny Rice-flower (Pimelea spinescens subsp. spinescens)*. Commonwealth of Australia, Canberra, <http://www.environment.gov.au/epbc/publications/pubs/spiny-rice-flower.pdf>.

Department of Sustainability, Environment, Water, Population and Communities (2011a) *Traveston Crossing Dam*. Commonwealth of Australia, Canberra, <http://www.environment.gov.au/cgi-bin/epbc/epbc_ap.pl?name=current_referral_detail&proposal_id=3150>.

Department of Sustainability, Environment, Water, Population and Communities (2011b) Development of a Model Code of Practice and Standard Operating Procedures for the Humane Capture, Handling or Destruction of Feral Animals in Australia, <http://www.environment.gov.au/biodiversity/invasive/publications/humane-control.html>.

Department of Sustainability, Environment, Water, Population and Communities (2011c) *Collaborative Australia Protected Area Database*. Commonwealth of Australia, Canberra, <http://www.environment.gov.au/parks/nrs/science/capad/index.html>.

Department of Sustainability, Environment, Water, Population and Communities (2011d) *Survey Guidelines for Australian Threatened Reptiles*. Commonwealth of Australia, Canberra, <http://www.environment.gov.au/epbc/publications/threatened-reptiles.html>.

Dignan P and Bren L (2003) A study of the effect of logging on the understorey light environment in riparian buffer strips in a south-east Australia forest. *Forest Ecology and Management* **172**, 161–172.

Dique DS, Thompson J, Preece HJ, Penfold GC, de Villiers DL and Leslie RS (2003) Koala mortality on roads in south-east Queensland: the koala speed-zone trial. *Wildlife Research* **30**, 419–426.

Dodd CK Jr, Barichivich WJ and Smith LL (2004) Effectiveness of a barrier wall and culverts in reducing wildlife mortality on a heavily traveled highway in Florida. *Biological Conservation* **118**, 619–631.

Donato D (1999) *Best Practice Guidelines for Reducing Impacts of Tailings Storage Facilities on Avian Wildlife in the Northern Territory.* Northern Territory Department of Mines and Energy, Darwin, <http://www.nt.gov.au/d/Minerals_Energy/Content/File/Forms_Guidelines/AVIAN_WILDLIFE_GUIDELINES.pdf>.

Donato DB, Nichols O, Possingham H, Moore M, Ricci PF and Noller BN (2007) A critical review of the effects of gold cyanide-bearing tailings solutions on wildlife. *Environment International* **33**, 974–984.

Dowle M and Deane EM (2009) Attitudes to native bandicoots in an urban environment. *European Journal of Wildlife Research* **55**, 45–52.

Downes SJ, Handasyde KA and Elgar MA (1997) The use of corridors by mammals in fragmented Australian eucalypt forests. *Conservation Biology* **11**, 718–726.

Drewitt AL and Langston RHW (2008) Collision effects of wind-power generators and other obstacles on birds. *Annals of the New York Academy of Sciences* **1134**, 233–266.

Driscoll DA, Lindenmayer DB, Bennett AF, Bode M, Bradstock RA, Cary GJ et al. (2010) Fire management for biodiversity conservation: key research questions and our capacity to answer them. *Biological Conservation* **143**, 1928–1939.

Ducummon SL (1999) The North American bats and mines project: a cooperative approach for integrating bat conservation and mine-land reclamation. *Proceedings of the Twenty-third Annual British Columbia Mine Reclamation Symposium – Mine Decommissioning,* 20–23 September Kamloops, Canada. pp. 10–21. British Columbia Technical and Research Committee on Reclamation, Canada.

Durant R, Luck GW and Matthews A (2009) Nest-box use by arboreal mammals in a peri-urban landscape. *Wildlife Research* **36**, 565–573.

Dwyer PD (1969) Population Ranges of *Miniopterus schreibersii* Chiroptera in Southeastern Australia. *Australian Journal of Zoology* **17**, 665–686.

Eamus D, Hatton T, Cook P and Colvin C (2006) *Ecohydrology: Vegetation Function, Water and Resource Management.* CSIRO Publishing, Melbourne.

Ecologically Sustainable Development Steering Committee (1992) *Australia's National Strategy for Ecologically Sustainable Development 1992.* Endorsed by the Council of Australian Governments. Australian Government Publishing Services, Canberra.

Edgar JP, Appleby RG and Jones DN (2007) Efficacy of an ultrasonic device as a deterrent to dingoes (*Canis lupus dingo*): a preliminary investigation. *Journal of Ethology* **25**, 209–213.

Ehmann H and Cogger H (1985) Australia's Endangered Herpetofauna: A Review of Criteria and Policies in Biology of Australasian Frogs and Reptiles. (Eds G Griss, R Shine and H Ehmann) Royal Zoological Society of New South Wales, Sydney.

Eigenbrod F, Hecnar SJ and Fahrig L (2009) Quantifying the road-effect zone: threshold effects of a motorway on anuran populations in Ontario, Canada. *Ecology and Society* **14,** [online] <http://www.ecologyandsociety.org/vol14/iss1/art24/>.

Eleuterio AA and Perez-Salicrup D (2009) Transplanting tree ferns to promote their conservation in Mexico. *American Fern Journal* **99**, 279–291.

Elliott M and Thomas I (2009) Environmental Impact Assessment in Australia – Theory and Practice. The Federation Press, Sydney.

Ensbey R (2009) Noxious and Environmental Weed Control Handbook: A Guide to Weed Control in Non-crop, Aquatic and Bushland Situations, Fourth Edition. Department of Industry and Investment, Sydney.

Environment Australia (2003) *Recovery Plan for Marine Turtles in Australia*. Commonwealth of Australia, Canberra.

Environmental Institute of Australia and New Zealand Ecology (2010) *Ecological Impact Assessment Guidelines – First Working Draft*. Prepared by Environmental Institute of Australia and New Zealand Ecology, Melbourne, <http://www.eianz.org/aboutus/ecology/ecology-sis>

Ermert S (2001) *Gardner's Companion to Weeds, Second Edition*. New Holland Publishers, Sydney.

Fahrig L, Pedlar JH, Pope SE, Taylor PD and Wegner JF (1995) Effect of road traffic on amphibian density. *Biological Conservation* **73**, 177–182.

Fahselt D (2007) Is transplanting an effective means of preserving vegetation? *Canadian Journal of Botany* **85**, 1007–1017.

Fensham RJ and Fairfax RJ (2007) Water-remoteness for grazing relief in Australian aridlands. *Biodiversity Conservation* **141**, 1447–1460.

Fernandez-Juricic E, Venier MP, Renison D and Blumstein DT (2004) Sensitivity of wildlife to spatial patterns of recreationist behaviour: a critical assessment of minimum approaching distances and buffer areas for grassland birds. *Biological Conservation* **125**, 225–235.

Ficetola GF, Sacchi R, Scali S, Gentilli A, De Bernardi F and Galeotti P (2006) Vertebrates respond differently to human disturbance: implications for the use of a focal species approach. *Acta Oecologia* **31**, 109–118.

Field SA, O'Connor PJ, Tyre AJ and Possingham HP (2007) Making monitoring meaningful. *Austral Ecology* **32**, 485–491.

Filip GM, Parks CG, Baker FA and Daniels SE (2004) Technical note – artificial inoculation of decay fungi into Douglas-Fir with rifle or shotgun to produce wildlife trees in western Oregon. *Western Journal of Applied Forestry* **19**, 211–215.

Findlay CS and Bourdages J (2000) Response time of wetland biodiversity to road construction on adjacent lands. *Conservation Biology* **14**, 86–94.

Fischer J and Lindenmayer DB (2000) An assessment of the published results of animal relocations. *Biological Conservation* **96**, 1–11.

Fischer J and Lindenmayer DB (2002) The conservation value of paddock trees for birds in a variegated landscape in southern New South Wales. 1. Species composition and site occupancy patterns. *Biodiversity and Conservation* **11**, 807–832.

Fischer J, Fazey J, Briese R and Lindenmayer DB (2005) Making the matrix matter: challenges in Australian grazing landscapes. *Biodiversity and Conservation* **14**, 561–578.

Fischer J, Scott J and Law BS (2009) The disproportional value of scattered trees. *Biological Conservation* **143**, 1564–1567.

Fisher AM and Goldney DC (1997) Use by birds of riparian vegetation in an extensively fragmented landscape. *Pacific Conservation Biology* **3**, 275–288.

FitzGibbon SI and Jones DN (2006) A community-based wildlife survey: the knowledge and attitudes of residents of suburban Brisbane, with a focus on bandicoots. *Wildlife Research* **33**, 233–241.

Flaquer C, Torre I and Ruiz-Jarillo R (2006) The value of bat-boxes in the conservation of *Pipistrellus pygmaeus* in wetland rice paddies. *Biological Conservation* **128**, 223–230.

Ford AT, Clevenger AP and Bennett A (2009) Comparison of methods of monitoring wildlife crossing-structures on highways. *Journal of Wildlife Management* **73**, 1213–1222.

Forman RTT and Alexander LE (1998) Roads and their major ecological effects. *Annual Review of Ecology and Systematics* **29**, 207–231.

Forman RTT, Sperling D, Bissonette JA, Clevenger AP, Cutshall CD, Dale VH et al. (2003) Road Ecology; Science and Solutions. Island Press, Washington DC.

Fowler C (2008) The Svalbard Seed Vault and crop security. *Biosciences* **58**, 190–191.

Frank KD (2006) Effect of artificial night lighting on moths. In *Ecological Consequences of Artificial Night Lighting*. (Eds C Rich and T Longcore) pp. 305–344. Island Press, Washington DC.

Franks A and Franks S (2003) *Nest Boxes for Wildlife: A Practical Guide*. Bloomings Books, Melbourne.

Frazer DS and Petit S (2007) Use of *Xanthorrhoea semiplana* (grass-trees) for refuge by *Rattus fuscipes* (southern bush rat). *Wildlife Research* **34**, 379–386.

Friend DA and Kemp DR (2000) Grazing management methods. In *Australian Weed Management Systems*. (Ed. MB Sindel) pp. 139–159. RG and FJ Richardson. Meredith, Victoria.

Gardyne W (1995) *Cleveland-Redland Bay Road Duplication-Koala Issues*. Unpublished report. Queensland Department of Transport, Transport Technology Division.

Garnett ST, Pedler LP and Crowley GM (1999) The breeding biology of the Glossy Black-cockatoo *Calyptorhynchus lathami* on Kangaroo Island, South Australia. *Emu* **99**, 262–279.

Geering A, Agnew L and Harding S (2007) *Shorebirds of Australia*. CSIRO Publishing, Melbourne.

Gentle CB and Duggin JA (1997) Allelopathy as a competitive strategy in persistent thickets of *Lantana camara* L. in three Australian forest communities. *Plant Ecology* **132**, 85–95.

Gibbons P and Lindenmayer D (2002) *Tree Hollows and Wildlife Conservation in Australia*. CSIRO Publishing, Melbourne.

Gibbons P and Lindenmayer DB (2007) Offsets for land clearing: no net loss or the tail wagging the dog? *Ecological Management and Restoration* **8**, 26–31.

Gibbons P, Lindenmayer DB, Barry SC and Tanton MT (2002) Hollow selection by vertebrate fauna in forests of southeastern Australia and implications for forest management. *Biological Conservation* **103**, 1–12.

Gibbons P, Lindenmayer DB, Fischer J, Manning AD, Weinderg A, Seddon J, Ryan P and Barrett G (2008) The future of scattered trees in agricultural landscapes. *Conservation Biology* **22**, 1309–1319.

Gilbert-Norton L, Wilson R, Stevens JR and Beard KH (2010) A meta-analytic review of corridor effectiveness. *Conservation Biology* **24**, 660–668.

Gilsdorf JM, Hygnstrom SE, VerCauteren KC, Clements GM, Blankenship EE and Engeman RM (2004) Evaluation of a deer-activated bio-acoustic frightening device for reducing deer damage in cornfields. *Wildlife Society Bulletin* **32**, 515–523.

Glista DJ, DeVault TL and DeWoody JA (2008) Vertebrate road mortality predominantly impacts amphibians. *Herpetological Conservation and Biology* **3**, 77–87.

Glista DJ, DeVault TL and DeWoody JA (2009) A review of mitigation measures for reducing wildlife mortality on roadways. *Landscape and Urban Planning* **91**, 1–7.

Goldingay RL (2008) Conservation of the endangered Green and Golden Bell Frog: what contribution has ecological research made since 1996? *Australian Zoologist* **34**, 334–349.

Goldingay RL (2009) Characteristics of tree hollows used by Australian birds and bats. *Wildlife Research* **36**, 394–409.

Goldingay RL and Newell DA (2000) Experimental rock outcrops reveal continuing habitat disturbance for an endangered Australian snake. *Conservation Biology* **14**, 1908–1912.

Goldingay RL and Stevens JR (2009) Use of artificial tree hollows by Australian birds and bats. *Wildlife Research* **36**, 81–97.Goldingay RL and Taylor BD (2006) How many frogs are killed on a road in north-east New South Wales? *Australian Zoologist* **33**, 332–336.

Goldingay RL and Taylor BD (2009) Gliding performance and its relevance to gap crossing by the squirrel glider (*Petaurus norfolcensis*). *Australian Journal of Zoology* **57**, 99–104.

Goldingay RL, Grimson MJ and Smith GC (2007) Do feathertail gliders show a preference for nest box design? *Wildlife Research* **34**, 484–490.

Goldingay RL, Taylor BD and Ball T (2011) Wooden poles can provide habitat connectivity for a gliding mammal. *Australian Mammalogy* **33**, 36–43.

Goldney D, Bauer J, Bryant H, Hodgkins D and Watson G (1996) Winning battles but losing the war: the education marketing imperative. In *Nature Conservation 4: The Role of Networks*. (Eds DA Saunders, JL Craig and EM Mattiske) pp. 574–588. Surrey Beatty & Sons, Chipping Norton, UK.

Goosem M (2000) Effects of tropical rainforest roads on small mammals: edge changes in community composition. *Wildlife Research* **27**, 151–163.

Goosem M (2005) *Wildlife Surveillance Assessment Compton Road Upgrade 2005: Research Report*. Report to the Brisbane City Council. Cooperative Research Centre for Tropical Rainforest Ecology and Management. Rainforest CRC, Cairns, <http://www.jcu.edu.au/rainforest/publications/compton_upgrade.pdf>.

Goosem M, Weston N and Bushnell S (2005) Effectiveness of rope bridge arboreal overpasses and faunal underpasses in providing connectivity for rainforest fauna. In *Proceedings of the 2005 International Conference on Ecology and Transportation*. 29 August –2 September 2005(Eds CL Irwin, P Garrett and KP McDermott) pp. 304–316. Center for Transportation and the Environment, North Carolina State University, Raleigh, NC.

Goosem M, Izumi Y and Turton S (2001) Will underpasses below roads restore habitat connectivity for tropical rainforest fauna? *Ecological Management and Restoration* **2**, 196–202.

Gordon KM, McKinstry MC and Anderson SH (2004) Motorist response to a deer-sensing warning system. *Wildlife Society Bulletin* **32**, 565–573.

Gosper CR (2004) Consequences of weed invasion and control on plant-bird interactions and bird communities. PhD dissertation. Department of Biological Sciences, University of Wollongong, Australia.

Gosper CR and Vivian-Smith G (2009) Approaches to selecting native plant replacements for fleshy-fruited invasive species. *Restoration Ecology* **17**, 196–204.

Grant C and Koch J (2007) Decommissioning Western Australia's first bauxite mine: co-evolving vegetation restoration techniques and targets. *Ecological Management and Restoration* **8**, 92–105.

Grant P (2003) Habitat Garden: Attracting Wildlife to Your Garden. ABC Books, Sydney.

Green R and Giese M (2004) Negative effects of wildlife tourism on wildlife. In *Wildlife Tourism-Impacts, Management and Planning*. (Ed. K Higginbottom) pp. 91–93. Common Ground Publishing Pty Ltd and Cooperative Research Centre for Sustainable Tourism.

Greening Australia (2011) *Florabank Network*. Greening Australia, <http://www.florabank.org.au/default.asp?V_DOC_ID=821>.

Grilo C, Bissonette JA and Santos-Reis M (2008) Response of carnivores to existing highway culverts and underpasses: implications for road planning and mitigation. *Biodiversity Conservation* **17**, 1685–1699.

Groffman PM, Baron JS, Blett T, Gold AJ, Goodman I, Gunderson LH *et al.* (2006) Ecological thresholds: the key to successful environmental management or an important concept with no practical application? *Ecosystems* **9**, 1–13.

Gunn B (2001) Australian Tree Seed Centre Operations Manual. CSIRO, Canberra.

Hall LS and Richards GC (1998) Issues concerning bat distribution on mine sites in Australia. In *Fauna Habitat Reconstruction after Mining Workshop*. (Eds CJ Asher and LC Bell) pp. 125–137. Australian Centre for Mining Environmental Research, Kenmore.

Hall LS, Richards G, McKenzie N and Dunlop N (1997) The importance of abandoned mines as habitat for bats. In *Conservation Outside Nature Reserves*. (Eds P Hales and D Lamb) pp. 326–333. Centre for Conservation Biology, The University of Queensland, Brisbane.

Hamer AJ and Mahony MJ (2010) Rapid turnover in site occupancy of a pond-breeding frog demonstrates the need for landscape-level management. *Wetlands* **30**, 287–299.

Hamer AJ and Organ AK (2008) Aspects of the ecology and conservation of the growling grass frog *Litoria raniformis* in an urban-fringe environment, southern Victoria. *Australian Zoologist* **34**, 393–407.

Hamer AJ, Lane SJ and Mahony MJ (2002) Management of freshwater wetlands for the endangered green and golden bell frog (*Litoria aurea*): roles of habitat determinants and space. *Biological Conservation* **106**, 413–424.

Hamer AJ, Lane SJ and Mahony MJ (2008) Movement patterns of adult Green and Golden Bell Frogs *Litoria aurea* and the implications for conservation management. *Journal of Herpetology* **42**, 397–407.

Hanger J and Nottidge B (2009) Queensland Code of Practice for the Welfare of Wild Animals Affected by Land-Clearing and Other Habitat Impacts and Wildlife Spotter/Catchers – Draft. Australian Wildlife Hospital (A Division of Australian Zoo Wildlife Warriors Worldwide Ltd).

Hardy A, Lee S and Al-Kaisy AF (2006) Effectiveness of animal advisory messages on dynamic message signs as a speed reduction tool – case study in rural Montana. In *Transportation Research Record Series* **1973**, 64–72.

Harness RE and Carlton R (2001) New solutions for bird collision and electrocution outage problems. In *Proceedings of the 2001 IEEE Power Engineering Society Winter Meeting*. 28 January–1 February 2001, Columbus. pp. 341–354. IEEE Power Engineering Society, New York.

Harris IM, Mills HR and Bencini R (2010) Multiple individual southern brown bandicoots (*Isoodon obesulus fusciventer*) and foxes (*Vulpes vulpes*) use underpasses installed at a new highway in Perth, Western Australia. *Wildlife Research* **37**, 127–133.

Harris LD and Scheck J (1991) From implications to applications: the dispersal corridor principle applied to the conservation of biological diversity. In *Nature Conservation 2: The Role of Corridors*. (Eds DA Saunders and RJ Hobbs) pp. 189–220. Surrey Beatty & Sons, Chipping Norton, UK.

Hayes IF and Goldingay RL (2009) Use of fauna road-crossing structures in north-eastern New South Wales. *Australian Mammalogy* **31**, 89–95.

Hazell D (2003) Frog ecology in modified Australian landscapes: a review. *Wildlife Research* **30**, 193–205.

Hazell D, Cunningham R, Lindenmayer D, Mackey B and Osbourne W (2001) Use of farm dams as frog habitat in an Australian agricultural landscape: factors affecting species richness and distribution. *Biological Conservation* **102**, 155–169.

Hazell D, Hero J, Lindenmayer D and Cunningham R (2004) A comparison of constructed and natural habitat for frog conservation in an Australian agricultural landscape. *Biological Conservation* **119**, 61–71.

Heard GW, Scroggie MP and Clemann N (2010) *Guidelines for Managing the Endangered Growling Grass Frog in Urbanising Landscapes*. Arthur Rylah Institute for Environmental Research Technical Report Series No. 208. Department of Sustainability and Environment, Heidelberg, Victoria.

Hellmund PC and Smith DS (2006) Designing Greenways – Sustainable Landscapes for Nature and People. Island Press, Washington DC.

Hels T and Buchwald E (2001) The effect of road kills on amphibian populations. In *Proceedings of the 2001 International Conference on Ecology and Transportation*. 2–24 September,

Keystone, Colorado (Eds CL Irwin, P Garrett and KP McDermott) pp. 25–42. Center for Transportation and the Environment, North Carolina State University, Raleigh, NC.

Herath DN and Lamont BB (2009) Persistence of resprouting species after fire in natural and post-mine restored shrublands in southwestern Australia. *Applied Vegetation Studies* **12**, 451–458.

Hess GR and Fischer RA (2001) Communicating clearly about conservation corridors. *Landscape and Urban Planning* **55**, 195–208.

Hezri AA and Dovers SR (2009) Australia's indicator-based sustainability assessments and public policy. *The Australian Journal of Public Administration* **68**, 303–318.

Hill D and Edquist N (1982) *Wildlife and Farm Dams.* Fisheries and Wildlife Division of the Soil Conservation Authority, Melbourne.

Hilty JA, Lidicker WZ Jr and Merenlender AM (2006) Corridor Ecology – The Science and Practice of Linking Landscapes for Biodiversity Conservation. Island Press, Washington DC.

Hobbs RJ (1992) The role of corridors in conservation: solution or bandwagon? *Tree* **7**, 389–392.

Hobbs RJ (1993) Can revegetation assist in the conservation of biodiversity in agricultural areas? *Pacific Conservation Biology* **1**, 29–38.

Hobbs RJ and Humphries SE (1995) An integrated approach to the ecology and management of plant invasions. *Conservation Biology* **9**, 761–770.

Hobbs RJ and Yates CJ (2003) Impacts of ecosystem fragmentation on plant populations: generalising the idiosyncratic. *Australian Journal of Botany* **51**, 471–488.

Hobbs RJ, Hallett LM, Ehrlich PR and Mooney HA (2011) Intervention ecology: applying ecological science in the twenty-first century. *BioScience* **61**, 442–450.

Hobday AJ (2010) Nighttime driver detection distances for Tasmanian fauna: informing speed limits to reduce roadkill. *Wildlife Research* **37**, 265–272.

Hobday AJ and Minstrell ML (2008) Distribution and abundance of roadkill on Tasmanian highways: human management options. *Wildlife Research* **35**, 712–726.

Homan P (2000) Excluding the common myna *Acridotheres tristis* from artificial next boxes using a baffle. *The Victorian Naturalist* **117**, 75.

Horn JW, Arnett EB, Jensen M and Kunz TH (2008) *Testing the Effectiveness of an Experimental Acoustic Bat Deterrent at the Maple Ridge Wind Farm.* Report prepared for the Bats and Wind Energy Cooperative and Bat Conservation International, Austin, TX.

Houston CS and Scott F (1992) The effect of man-made platforms on osprey reproduction at Loon Lake, Saskatchewan. *Journal of Raptor Research* **26**, 152–158.

Hoyle SD, Horsup AB, Johnson CN, Crossman DG and McCallum (1995) Live trapping of the northern hairy-nosed wombat (*Lasiorhinus krefftii*): population-size estimates and effect on individuals. *Wildlife Research* **22**, 741–755.

Huber PR, Greco SE and Thorne JH (2010) Spatial scale effects on conservation network design: tradeoffs and omissions in regional versus local scale planning. *Landscape Ecology* **25**, 683–695.

Huijser MP, McGowen P, Fuller J, Hardy A, Kociolek A, Clevenger AP, Smith D and Ament R (2008) *Wildlife-Vehicle Collision Reduction Study: Report to Congress.* US Department of Transport Federal Highway Administration. Washington DC.

Huijser MP, Duffield JW, Clevenger AP, Ament RJ and McGowen PT (2009) Cost-benefit analysis of mitigation measures aimed at reducing collisions with large ungulates in the United States and Canada: a decision support tool. *Ecology and Society* **14**, 15 [online], <http://www.ecologyandsociety.org/vol14/iss2/art15/>.

Huijser MP, Galarus DE and Kociolek AV (2010) Current and developing technologies in highway-wildlife mitigation. In *Safe Passages: Highways, Wildlife and Habitat Connectivity.* (Eds JP Beckmann, AP Clevenger, MP Huijser and JA Hilty) pp. 309–321. Island Press, Washington DC.

Hull CL and Muir S (2010) Search areas for monitoring bird and bat carcasses at wind farms using a Monte-Carlo model. *Australasian Journal of Environmental Management* **17**, 77–87.

Hutson M, Mickleburgh SP and Racey PA (2001) *Microchiropteran Bats: Global Status Survey and Conservation Action Plan*. Information Press, Oxford, UK.

Hylander K, Nilsson C and Gothner T (2004) Effects of buffer-strip retention and clearcutting on land snails in boreal riparian forests. *Conservation Biology* **18**, 1052–1062.

IEEE Task Force on Reducing Bird Related Power Outages (2004) Preventive measures to reduce bird-related power outages – part 1: electrocution and collision. *IEEE Transactions on Power Delivery* **19**, 1843–1847.

International Bird Strike Committee (2006) *Recommended Practices No.1 Standards for Aerodrome Bird/Wildlife Control*. International Bird Strike Committee, <http://www.int-birdstrike.org/Best_Practice.htm>.

International Union for Conservation of Nature (1987) *Translocation of Living Organisms International Union for Conservation of Nature Position Statement*. Prepared by the Species Survival Commission in collaboration with the Commission on Ecology, and the Commission on Environmental Policy, Law and Administration. International Union for Conservation of Nature, Gland, Switzerland.

International Union for Conservation of Nature (2009) *International Union for Conservation of Nature Protected Areas Categories System*. International Union for Conservation of Nature, Gland, Switzerland, <http://www.iucn.org/about/work/programs/pa/pa_products/wcpa_categories>.

Iuell B, Bekker GJ, Cuperus R, Dufek J, Fry G, Hicks C et al. (2003) Wildlife and Traffic: A European Handbook for Identifying Conflicts and Designing Solutions – Cooperation in the field of Scientific and Technical Research (COST) 341 Habitat Fragmentation due to Transportation Infrastructure. Infra Eco Network Europe, Uppsala, Sweden <http://forum.iene.info/cost-341/european-handbook>.

Jaeger JAG, Bowman J, Brennan J, Fahrig L, Bert D, Bouchard J et al. (2005) Predicting when animal populations are at risk from roads: an interactive model of road avoidance behaviour. *Ecological Modelling* **185**, 329–348.

Jaeger JAG and Fahrig L (2004) Effects of road fencing on population persistence. *Conservation Biology* **18**, 1651–1657.

James AI and Eldridge DJ (2007) Reintroduction of fossorial native mammals and potential impacts on ecosystem processes in an Australian desert landscape. *Biological Conservation* **138**, 351–359.

Janss GFE (2000) Avian mortality from power lines: a morphologic approach of a species-specific mortality. *Biological Conservation* **95**, 353–359.

Janss GFE, Lazo A and Ferrer M (1999) Use of raptor models to reduce avian collisions with powerlines. *Journal of Raptor Research* **33**, 154–159.

Jenkins AR, Smallie JJ and Diamond M (2010) Avian collisions with power lines: a global review of causes and mitigation with a South African perspective. *Bird Conservation International* **20**, 263–278.

Jones DN and Bond ARF (2010) Road barrier effect on small birds removed by vegetated overpass in South East Queensland. *Ecological Management and Restoration* **11**, 65–67.

Jones DN, Bakker M, Bichet O, Coutts R and Wearing T (2011) Restoring habitat connectivity over the road: vegetation on a fauna land-bridge in south-east Queensland. *Ecological Management and Restoration* **12**, 76–79.

Jones J and Francis CM (2003) The effects of light characteristics on avian mortality at lighthouses. *Journal of Avian Biology* **34**, 328–333.

Jones ME (2000) Road upgrade, road mortality and remedial measures: impacts on a population of eastern quolls and Tasmanian devils. *Wildlife Research* **27**, 289–296.

Jotikapukkana S, Berg A and Pattanavibool A (2010) Wildlife and human use of buffer-zone areas in a wildlife sanctuary. *Wildlife Research* **37**, 466–474.

Julien MH (1992) Biological Control of Weeds. A World Catalogue of Agents and their Target Weeds. Commonwealth Agricultural Bureau, Wallingford, UK.

Kaczmarczyk A, Turner SR, Bunn E, Mancera RL and Dixon KW (2011) Cryopreservation of threatened native Australian species—what have we learned and where to from here? *In Vitro Cellular and Developmental Biology - Plant*. **47**, 17–25.

Keeley BW and Tuttle MD (1999) *Bats in American Bridges*. Bat Conservation International, Inc., Austin, Texas, <www.batcon.org>.

Keydoszius JR, Cox SK Jr, Haque MB, Milkhailova E, Post CJ, Stringer WC and Schlautman MA (2007) Historical land use and soil analysis guiding corridor landscape design. *Urban Ecosystems* **10**, 53–72.

Kim DH, Chavez-Ramirez F and Slack RD (2003) Effects of artificial perches and interspecific interactions on patch use by wintering raptors. *Canadian Journal of Zoology* **81**, 2038–2047.

Kimber SL, Bennett AF and Ryan PA (1999) *Revegetation and Wildlife: What Do We Know about Revegetation and Wildlife Conservation in Australia?* Environment Australia, <http://www.environment.gov.au/land/publications/wildlife.html>.

Kinross C and Nicol H (2008) Responses to birds to the characteristics of farm windbreaks in central New South Wales, Australia. *Emu* **108**, 139–152.

Kirchner F, Ferdy JB, Andalo, C, Colas B and Moret J (2003) Role of corridors in plant dispersal: an example with the Endangered *Ranunculus nodiflorus*. *Conservation Biology* **2**, 401–410.

Klem D (2009) Preventing bird-window collisions. *Wilson Journal of Ornithology* **121**, 314–321.

Klem D, Farmer CJ, Delacretaz N, Gelb Y and Saenger PG (2009) Architectural and landscape risk factors associated with bird-glass collision in an urban environment. *The Wilson Journal of Ornithology* **12**, 126–134.

Klocker U, Croft DB and Ramp D (2006) Frequency and causes of kangaroo-vehicle collisions on an Australian outback highway. *Wildlife Research* **33**, 5–15.

Knapp D and Yang L (2002) A phenomenological analysis of long-term recollections of an interpretive program. *Journal of Interpretation Research* **7**, 7–17.

Koch JM and Hobbs RJ (2007) Synthesis: is Alcoa successfully restoring a jarrah forest ecosystem after bauxite mining in Western Australia. *Restoration Ecology* **15**, 137–144.

Koenig J, Shine R and Shea G (2001) The ecology of an Australian reptile icon: how do blue-tongued lizards (*Tiliqua scincoides*) survive in suburbia? *Wildlife Research* **28**, 215–227.

Krebs EA (1998) Breeding biology of crimson rosellas (*Platycercus elegans*) on Black Mountain, Australian Capital Territory. *Australian Journal of Zoology* **46**, 119–136.

Kristoffersen P, Rask AM and Larsen SU (2008) Non-chemical weed control on traffic islands: a comparison of the efficacy of five weed control techniques. *Weed Research* **48**, 124–130.

Kubik R (2007) How to Build and Repair Fences and Gates. Voyageur Press, St Paul, Minnesota.

Laliberté B (1997) Botanic Garden Seed Banks/Genebanks Worldwide, their Facilities, Collections and Networks. BGCNews, Richmond, UK, <http://www.bgci.org/worldwide/article/0032/>.

Lalo J (1987) The problem of roadkill. *American Forests* **50**, 50–52.

Lamont BB, Wittkuhn R and Korczyskyj D (2004) Turner review no. 8, ecology and ecophysiology of grasstrees. *Australian Journal of Botany* **52**, 561–582.

Landsberg J, James CD, Morton SR, Hobbs TJ, Stol J, Drew A and Tongway H (1997) *The Effects of Artificial Sources of Water on Rangeland Biodiversity.* Panther Publishing and Printing, Canberra.

Law BS and Chidel M (2007) Bats under a hot tin roof: comparing the microclimate of eastern cave bat (*Vesadelus troughtoni*) roosts in a shed and cave overhangs. *Australian Journal of Zoology* **55**, 49–55.

Leach GJ (1994) Effects of dam size on waterbirds at farm dams in south-east Queensland. *Corella* **18**, 77–82.

Lee C, Henshall JM, Wark TJ, Crossman CC, Reed MT, Brewer HG, O'Grady JO and Fisher AD (2009) Associative learning by cattle to enable effective and ethical virtual fences. *Applied Animal Behaviour Science* **119**, 15–22.

Lee J, Finn H and Calver MC (2010) Mine-site revegetation monitoring detects feeding by threatened black-cockatoos within 8 years. *Ecological Management and Restoration* **11**, 141–143.

Lemckert F and Brassil T (2000) Movements and habitat use of the endangered giant barred rover frog (*Mixophyes iterates*) and the implications for its conservation in timber production forests. *Biological Conservation* **96**, 177–184.

Lesbarrères D, Lodé T and Merilä J (2004) What type of amphibian tunnel could reduce road kills? *Oryx* **38**, 220–223.

Lettink M and Cree A (2007) Relative use of three types of artificial retreats by terrestrial lizards in grazed coastal shrubland, New Zealand. *Applied Herpetology* **4**, 227–243.

Lewis B (2002) Farm Dams: Planning, Construction and Maintenance. Landlinks Press, Melbourne.

Lewis SE (1995) Roost fidelity of bats: a review. *Journal of Mammalogy* **76**, 481–496.

Ley A and Tynan R (2008) Birds Killed by Fences in Diamantina National Park, Queensland. *Australian Field Ornithology* **26**, 96–98.

Lilith M, Calver M, Styles I and Garkaklis M (2006) Protecting wildlife from predation by owned domestic cats: application of a precautionary approach to the acceptability of proposed cat regulations. *Austral Ecology* **31**, 176–189.

Limpus CJ (2008) A biological review of Australian marine turtles. 2. Green Turtle Chelonia mydas (Linnaeus). The State of Queensland. Environmental Protection Agency.

Lindenmayer DB (2011) *What Makes a Good Farm for Wildlife?* CSIRO Publishing, Melbourne.

Lindenmayer DB and Burgman M (2005) *Practical Conservation Biology.* CSIRO Publishing, Melbourne.

Lindenmayer DB and Fischer J (2002) Sound science or social hook – a response to Brookers application of the focal species approach. *Landscape and Urban Planning* **62**, 149–158.

Lindenmayer DB and Fischer J (2006) Habitat Fragmentation and Landscape Change: An Ecological and Conservation Synthesis. CSIRO Publishing, Melbourne.

Lindenmayer DB and Likens GE (2010a) *Effective Ecological Monitoring.* CSIRO Publishing, Melbourne.

Lindenmayer DB and Likens GE (2010b) The science and application of ecological monitoring. *Biological Conservation* **143**, 1317–1328.

Lindenmayer DB, Claridge AW, Gilmore AM, Michael D and Lindenmayer BD (2002) The ecological roles of logs in Australian forests and the potential impacts of harvesting intensification on log-using biota. *Pacific Conservation Biology* **8**, 121–140.

Lindenmayer DB, Claridge A, Hazell D, Michael D, Crane M, MacGregor C and Cunningham R (2003) *Wildlife on Farms: How to Conserve Animals.* CSIRO Publishing, Melbourne.

Little SJ, Harcourt RG and Clevenger AP (2002) Do wildlife passages act as prey-traps? *Biological Conservation* **107**, 135–145.

Lloyd MV, Barnett G, Doherty MD, Jeffree RA, John J, Majer JD, Osborne JM and Nichols OG (2002) *Managing the Impacts of the Australian Minerals Industry on Biodiversity.* Australian Centre for Mining Environmental Research (ACMER), <http://www.acmer.uq.edu.au/publications/attachments/MMSDReportApril02.pdf>

Loney B and Hobbs RJ (1991) Management of vegetation corridors: maintenance, rehabilitation and establishment. In *Nature Conservation 2: The Role of Corridors.* (Eds DA Saunders and RJ Hobbs) pp. 241–246. Surrey Beatty & Sons, Chipping Norton, UK.

Long K and Robley A (2004) Cost Effective Feral Animal Exclusion Fencing for Areas of High Conservation Value in Australia. Commonwealth of Australia, Canberra.

Longcore T and Rich C (2004) Ecological light pollution. *Frontiers in Ecology and the Environment* **2**, 191–198.

Longcore T and Rich C (2006) Synthesis. In *Ecological Consequences of Artificial Night Lighting.* (Eds T Longcore and C Rich) pp. 413–426. Island Press, Washington DC.

Loud M (2000) Dispersal of soaring raptors using radio-controlled aircraft. In *Proceedings of the Bird Strike Committee – USA/Canada.* 8–10 August 2000, Minneapolis. Poster. p. 31. University of Nebraska – Lincoln.

Lowe I (2009) A Big Fix: Radical Solutions for Australia's Environmental Crisis. Black Inc., Melbourne.

Lumsden HG (1989) Test of nest box preferences of eastern bluebirds, *Sialia sialis*, and tree swallows, *Tachycineta bicolor. The Canadian Field Naturalist* **103**, 595–597.

Lynch JF and Saunders DA (1991) Responses of bird species to habitat fragmentation in the wheatbelt of Western Australia: interiors, edges and corridors. In *Nature Conservation 2: The Role of Corridors.* (Eds DA Saunders and RJ Hobbs) pp. 99–117. Surrey Beatty & Sons, Chipping Norton, UK.

Mackenzie BDE and Keith DA (2009) Adaptive management in practice: conservation of a threatened plant population. *Ecological Management and Restoration* **10**, 129–135.

Mac Nally R and Horrocks G (2007) Inducing whole-assemblage change by experimental manipulation of habitat structure. *Journal of Animal Ecology* **76**, 643–650.

Mac Nally R and Horrocks G (2008) Longer-term responses of a floodplain-dwelling marsupial to experimental manipulation of fallen timber loads. *Basic and Applied Ecology* **9**, 458–465.

Mac Nally R, Horrocks G and Pettifer L (2002) Experimental evidence for potential beneficial effects of fallen timber in forests. *Ecological Applications* **12**, 1588–1594.

Magnus Z, Kriwoken LK, Mooney NJ and Jones ME (2004) *Reducing the Incidence of Wildlife Roadkill: Improving the Visitor Experience in Tasmania.* Cooperative Research Centre for Sustainable Tourism. Gold Coast, Queensland.

Majer JD, Brennan KEC and Moir ML (2007) Invertebrate and the restoration of a forest ecosystem: 30 years of research following bauxite mining in Western Australia. *Restoration Ecology* **15**, 104–115.

Mann RM, Hyne RV, Choung CB and Wilson SP (2009) Amphibians and agricultural chemicals: review of the risks in a complex environment. *Environmental Pollution* **157**, 2903–2927.

Manning AD, Fischer J and Lindenmayer DB (2006) Scattered trees are keystone structures – implications for conservation. *Biological Conservation* **132**, 311–321.

Mansergh IM and Scotts DJ (1989) Habitat continuity and social organization of the mountain pygmy-possum restored by tunnel. *Journal of Wildlife Management* **53**, 701–707.

Maron M, Dunn PK, McAlpine CA and Apan A (2010) Can offsets really compensate for habitat removal? The case of the endangered red-tailed black-cockatoo. *Journal of Applied Ecology* **47**, 348–355.

Marquez-Ferrando R, Pleguezuelos JM, Santos X, Ontiveros D and Fernandez-Cardenete JR (2009) Recovering the reptile community after the mine-tailing accident of Aznalcollar (Southwestern Spain). *Restoration Ecology* **17**, 660–667.

Marsh RE, Erickson WA and Salmon TP (1991) *Bird Hazing and Frightening Methods and Techniques (with Emphasis on Containment Ponds).* Department of Wildlife and Fisheries Biology University of California, Davis.

Martin GR (2011) Understanding bird collisions with man-made objects: a sensory ecology approach. *Ibis* **153**, 239–254.

Martin GR and Shaw JM (2010) Bird collisions with power lines: failing to see the way ahead? *Biological Conservation* **143**, 2695–2702.

Martin LR and Hagar S (1990) Bird control on containment pond sites. In *Proceedings of the Fourteenth Vertebrate Pest Conference*, 6–8 March, Sacramento, California (Eds LR Davis and RE Marsh) pp. 307–310. University of California, Davis.

Mazerolle MJ (2004) Drainage ditches facilitate frog movements in a hostile landscape. *Landscape Ecology* **20**, 579–590.

McCall S C, McCarthy M A, van der Ree R, Harper M J, Cesarini S and Soanes K (2010) Evidence that a highway reduces apparent survival rates of squirrel gliders. *Ecology and Society* **15**, 27 [online]. <http://www.ecologyandsociety.org/vol15/iss3/art27/>.

McClanahan TR and Wolfe RW (1993) Accelerating forest succession in a fragmented landscape: the role of birds and perches. *Conservation Biology* **7**, 279–288.

McDonald T and Williams J (2009) A perspective on the evolving science and practice of ecological restoration in Australia. *Ecological Management and Restoration* **10**, 113–125.

McFadden M, Duffy S, Harlow P, Hobcroft D, Webb C and Ward-Fear G (2008) A review of the green and golden bell frog *Litoria aurea* breeding program at Taronga Zoo. *Australian Zoologist* **34**, 291–296.

McIntyre S, McIvor JG and Heard KM (Eds) (2002) *Managing and Conserving Grassy Woodlands.* CSIRO Publishing, Melbourne.

McIvor JG and McIntyre S (2002) Understanding grassy woodland ecosystems. In *Managing and Conserving Grassy Woodlands.* (Eds S McIntyre, JG McIvor and KM Heard) pp. 1–24. CSIRO Publishing, Melbourne.

McKergow LA, Prosser IP, Weaver DM, Grayson RB and Reed AEG (2006) Performance of grass and eucalyptus riparian buffers in a pasture catchment, Western Australia, part 2: water quality. *Hydrological Processes* **20**, 2327–2346.

McNeil R, Rodriguez JR and Ouellet H (1985) Bird mortality at a power transmission line in northeastern Venezuela. *Biological Conservation* **31**, 153–165.

Michael DR, Lunt ID and Robinson WA (2004) Enhancing fauna habitat in grazed native grasslands and woodlands: use of artificially placed log refuges by fauna. *Wildlife Research* **31**, 65–71.

Miller KK (2003) Public and stakeholder values of wildlife in Victoria, Australia. *Wildlife Research* **30**, 465–476.

Mitchell-Jones AJ (2004) Conserving and creating bat roosts. In *Bat Workers' Manual, Third Edition.* (Eds AJ Mitchell-Jones and AP McLeish) pp. 111–134. Joint Nature Conservation Committee, UK.

Moloney S and Vanderwoude C (2002) Red imported fire ants: a threat to eastern Australia's wildlife? *Ecological Management and Restoration* **3**, 167–175.

Montana Department of Transportation (2007) *Roadkill Carcass Composting Guidance Manual*. Montana Department of Transportation. Helena, Montana, <http://www.mdt. mt.gov/publications/docs/manuals/roadkill_composting.pdf >.

Mooney PA and Pedler LP (2005) Recovery Plan for the South Australian subspecies of the Glossy Black-cockatoo (Calyptorhynchus lathami halmaturinus): 2005–2010. Unpublished report to South Australian Department for Environment and Heritage, Adelaide.

Moseby KE and Read JL (2006) The efficacy of feral cat, fox and rabbit exclusion fence designs for threatened species protection. *Biological Conservation* **127**, 429–437.

Moseby KE, Hill BM and Read J (2009) Arid recovery – a comparison of reptile and small mammal populations inside and outside a large rabbit, cat and fox-proof enclosure in arid South Australia. *Austral Ecology* **34**, 156–169.

Muir GW (2008) Design of a movement corridor for the green and golden bell frog *Litoria aurea* at Sydney Olympic Park. *Australian Zoologist* **34**, 297–302.

Muirhead S, Blache D, Wykes B and Bencini R (2006) Roo-Guard® sound emitters are not effective at deterring tammar wallabies (*Macropus eugenii*) from a source of food. *Wildlife Research* **33**, 131–136.

Mukoyama M (2005) Bat conservation in the Tohoku area, northern Japanese main island. *The Australasian Bat Society Newsletter* **25**, 57–61.

Munro NT, Fischer J, Wood J and Lindenmayer DB (2009) Revegetation in agricultural areas: the development of structural complexity and floristic diversity. *Ecological Applications* **19**, 1197–1210.

National Biodiversity Strategy Review Task Group (2010) *Australia's Biodiversity Conservation Strategy 2010–2030*. Australian Government, Department of the Environment, Water, Heritage and the Arts, Canberra.

Nemtzov SC and Olsvig-Whittaker L (2003) The use of netting over fishponds as a hazard to waterbirds. *Waterbirds: The International Journal of Waterbird Biology* **26**, 416–423.

New South Wales Botanic Gardens Trust (2011) *New South Wales Seed Bank*. New South Wales Botanic Gardens Trust, Sydney, <http://www.rbgsyd.nsw.gov.au/science/Horticultural_ Research/nsw_seedbank>.

New South Wales Department of Environment and Climate Change (2008a) *Best Practice Guidelines Green and Golden Bell Frog Habitat*. New South Wales Department of Environment and Climate Change, Sydney.

New South Wales Department of Environment and Climate Change (2008b) *Hygiene Protocol for the Control of Disease in Frogs*. New South Wales Department of Environment and Climate Change, Sydney.

New South Wales Department of Land and Water Conservation (1998) *The Constructed Wetlands Manual – Volumes 1 and 2*. Department of Land and Water Conservation, Sydney.

New South Wales Government Roads and Traffic Authority (2009) *Media release: Osprey chicks in pole position*, <http://www.rta.nsw.gov.au/constructionmaintenance/downloads/ northcoast/osprey_mediarelease_oct09.pdf>.

New South Wales National Parks and Wildlife Service (2001a) *Strategy for the Conservation of Bats in Derelict Mines*. New South Wales National Parks and Wildlife Service, Hurstville, NSW.

New South Wales National Parks and Wildlife Service (2001b) Threatened Species Management Policy and Procedures Statement No 9. Policy for the Translocation of Threatened Fauna in New South Wales. New South Wales National Parks and Wildlife Service, Hurstville, NSW.

New South Wales Roads and Traffic Authority (2011) *Fencing (R201),* <http://www.rta.nsw.gov.au/doingbusinesswithus/designdocuments/modelroaddrawings/mrd_fencing.html>.

New South Wales Rural Fire Service (2006) Planning for Bushfire Protection – A Guide for Councils, Planners, Fire Authorities and Developers. New South Wales Rural Fire Service, Lidcombe, NSW.

New Zealand Plant Conservation Network (2011) *More Surprising Finds of* Fissidens berteroi *In Auckland City.* New Zealand, <http://www.nzpcn.org.nz/news_detail.asp?Status=3&ID=177>.

Nichols OG and Grant CD (2007) Vertebrate fauna recolonization of restored bauxite mines: key findings from almost 30 years of monitoring and research. *Restoration Ecology* **15**, 116–126.

Nichols OG and Nichols FM (2003) Long-term trends in faunal recolonization after bauxite mining in the jarrah forest of South-western Australia. *Restoration Ecology* **11**, 261–272.

Nichols OG, Grant C and Bell LC (2005) Developing Ecological Completion Criteria to Measure the Success of Forest and Woodland Establishment On Rehabilitated Mines in Australia. In *Proceedings from the National Meetings of the American Society of Mining and Reclamation,* 18–24 June, Breckenridge Colorado. American Society of Mining and Reclamation, Lexington, Kentucky.

Norman T, Finegan A and Lean B (1998) The role of fauna underpasses in New South Wales. In *Proceedings of the International Conference on Wildlife Ecology and Transportation,* 9–12 February, Fort Myers, Florida. (Eds GL Evink, P Garrett and D Zeigier) pp. 195–208. Center for Transportation and the Environment, Raleigh, North Carolina.

O'Meara J and Darcovich K (2008) Gambusia control through the manipulation of water levels in Narawang Wetland, Sydney Olympic Park 2003–2005. *Australian Zoologist* **34**, 285–290.

Offord CA and Meager PF (2009) Plant Germplasm Conservation in Australia – Strategies and Guidelines for Developing, Managing and Utilising Ex situ Collections. Australian Network for Plant Conservation Inc., Canberra.

Ogden LJE (1996) Collision Course: The Hazards of Lighted Structures and Windows to Migrating Birds. Technical Report. World Wildlife Fund Canada and the Fatal Light Awareness Program, Toronto, Canada.

Olsson MPO, Widen P and Larkin JL (2008) Effectiveness of a highway overpass to promote landscape connectivity and movement of moose and roe deer in Sweden. *Landscape and Urban Planning* **85**, 133–139.

Orams MB (2002) Feeding wildlife as a tourism attraction: a review of issues and impacts. *Tourism Management* **23**, 281–293.

Organ A (2004) *Pakenham Bypass: Growling Grass Frog* Litoria raniformis *2003/04 Survey, Pakenham and Surrounds Victoria.* Unpublished report prepared by Biosis Research Pty. Ltd. for VicRoads, Melbourne.

Organ A (2005) *Pakenham Bypass: Conservation Management Plan for the Growling Grass Frog* Litoria raniformis, *Pakenham, Victoria.* Unpublished report prepared by Biosis Research Pty. Ltd. for VicRoads, Melbourne.

Organ A and Hamer A (2006) *Growling Grass Frog* Litoria raniformis *Monitoring 2005/06, Pakenham Bypass, Pakenham, Victoria.* Unpublished report by Ecology Partners Pty Ltd. for VicRoads, Melbourne.

Padua SM (2010) Primate conservation: integrating communities through environmental education programs. *American Journal of Primatology* **75**, 450–453.

Parris KM (2006) Urban amphibian assemblages as metacommunities. *Journal of Animal Ecology* **75**, 757–764.

Parsons JM (1992) Australian Weed Control Handbook, Ninth Edition. Inkata Press, Melbourne.

Paton D and O'Connor J (Eds) (2010) The State of Australia's Birds 2009 – Restoring Woodland Habitats for Birds. Supplement to Wingspan **20**.

Pedler RD (2010) The impacts of abandoned mining shafts: fauna entrapment in opal prospecting shafts at Coober Pedy, South Australia. *Ecological Management and Restoration* **11**, 36–42.

Peel B (2010) *Rainforest Restoration Manual for South-eastern Australia*. CSIRO Publishing, Melbourne.

Penman TD, Muir GW, Magarey ER and Burns EL (2008) Impact of a Chytrid-related mortality event on a population of the green and golden bell frog *Litoria aurea. Australian Zoologist* **34**, 314–318.

Pennell CGL, Rolston MP, De Bonth A, Simpson WR and Hume DE (2010) Development of a bird-deterrent fungal endophyte in turf tall fescue. *New Zealand Journal of Agricultural Research* **53**, 145–150.

Perlman DL and Milder JC (2005) *Practical Ecology for Planners, Developers and Citizens*. Island Press, Washington DC.

Perry G, Buchanan BW, Fisher RN, Salmon M and Wise SE (2008) Effects of artificial night lighting on amphibians and reptiles in urban environments. In *Urban Herpetology*. (Eds JC Mitchell, RE Jung Brown and B Bartholomew) pp. 239–256. Salt Lake City, Utah.

Pfennigwerth S (2008) Minimising the Swift Parrot Collision Threat – Guidelines and Recommendations for Parrot-safe Building Design. World Wide Fund for Nature Australia, Sydney.

Pike DA, Croak BM, Webb JK and Shine R (2010) Subtle – but easily reversible – anthropogenic disturbance seriously degrades habitat quality for rock-dwelling reptiles. *Animal Conservation* **13**, 411–418.

Pincock S (2006) *Yucky Grass Keeps Birds Off Golf Courses*. Australian Broadcasting Corporation Science, <http://www.abc.net.au/science/articles/2006/08/08/1708055.htm>.

Pocock Z and Lawrence RE (2005) How far into a forest does the effect of a road extend? Defining road edge effect in eucalypt forests of South-Eastern Australia. In *Proceedings of the 2005 International Conference on Ecology and Transportation*. San Diego, California. 29 August–2 September (Eds CL Irwin, P Garrett and KP McDermott) pp. 397–405. Center for Transportation and the Environment, North Carolina State University, Raleigh, NC.

Pratley JE (2000) Tillage and other physical management methods. In *Australian Weed Management Systems*. (Ed. BM Sindel) pp. 105–122. RG and FJ Richardson. Meredith, Victoria.

Preston C (2000) Herbicide mode of action and herbicide resistance. In *Australian Weed Management Systems*. (Ed. BM Sindel) pp. 209–225. RG and FJ Richardson. Meredith, Victoria.

Prober SM and Smith FP (2009) Enhancing biodiversity persistence in intensively used agricultural landscapes: a synthesis of 30 years of research in the Western Australian wheatbelt. *Agriculture, Ecosystems and Environment* **132**, 173–191.

Puky M (2005) Amphibian road kills: a global perspective. UC Davis: Road Ecology Center. In *Proceedings of the 2005 International Conference on Ecology and Transportation*. San Diego, California. 29 August–2 September (Eds CL Irwin, P Garrett and KP McDermott) pp. 325–338. Center for Transportation and the Environment, North Carolina State University, Raleigh, NC.

Pulsford I, Worboys G, Gough J and Shepherd T (2003) A potential new continental scale conservation corridor for Australia – combining the Australian Alps and the Great Escarpment of eastern Australia conservation corridors. *Mountain Research and Development* **23**, 291–293.

Pyke GH and White AW (1996) Habitat requirements for the green and golden bell frog *Litoria aurea* (Anura: Hylidae). *Australian Zoologist* **30**, 224–232.

Queensland Department of Environment and Resource Management (2011) *Xstrata Reintroduction Project*, <http://www.derm.qld.gov.au/wildlife-ecosystems/wildlife/threatened_plants_and_animals/endangered/northern_hairynosed_wombat/xstrata_reintroduction_project.html>.

Queensland Department of Main Roads (2000) *Fauna Sensitive Road Design. Volume 1 – Past and Existing Practices*. Queensland Department of Main Roads, Planning, Design Environment Division, Brisbane.

Queensland Department of Transport and Main Roads (2010) *Fauna Sensitive Road Design Manual Volume 2*. Queensland Department of Main Roads, Planning, Design Environment Division, Brisbane.

Queensland Environmental Protection Agency (2008a) *Tree Clearing and Trimming – Koala Spotter Requirements*. Queensland Environmental Protection Agency, Brisbane, <http://www.derm.qld.gov.au/register/p02132aa.pdf>.

Queensland Environmental Protection Agency (2008b) *Nature Conservation (Koala) Conservation Plan 2006*. Queensland Environmental Protection Agency, Brisbane.

Queensland Government (2009) *Standard Drawings Roads Manual*. Issued by the Department of Transport and Main Roads Engineering and Technology, Brisbane.

Radford J and Bennett A (2010) Species-area relationships and woodland birds. In *The State of Australia's Birds 2009 – Restoring Woodland Habitats for Birds*. (Eds D Paton and J O'Connor) Supplement to *Wingspan* **20**.

Ramos Pereira MJR, Salgueiro P, Rodrigues L, Coelho MM and Palmeirim JM (2009) Population structure of a cave-dwelling bat, *Miniopterus schreibersii*: does it reflect history and social organisation. *Journal of Heredity* **100**, 533–544.

Ramp D and Croft DB (2006) Do wildlife warning reflectors elicit aversion in captive macropods? *Wildlife Research* **33**, 583–590.

Ramp D and Roger E (2008) Frequency of animal-vehicle collisions in New South Wales. In *Royal Zoological Society of New South Wales Forum Too Close for Comfort: Contentious Issues in Human-Wildlife Encounters*. 21 October, Mosman. (Ed. D Lunney) p. 118. Royal Zoological Society of New South Wales, Sydney.

Ramp D, Caldwell J, Edwards KA, Warton D and Croft DB (2005a) Modelling of wildlife fatality hotspots along the Snowy Mountain Highway in New South Wales, Australia. *Biological Conservation* **126**, 474–490.

Ramp D, Russell B and Croft D (2005b) Predator scent induces differing responses in two sympatric macropodids. *Australian Journal of Zoology* **53**, 73–78.

Ramp D, Wilson VK and Croft DB (2006) Assessing the impacts of roads in peri-urban reserves: Road-based fatalities and road usage by wildlife in the Royal National Park, New South Wales, Australia. *Biological Conservation* **129**, 348–359.

Rankmore BR, Griffiths AD, Woinarski JCZ, Bruce Lirrwa Ganambarr, Taylor R, Brennan K, Firestone K and Cardoso M (2009) *Island Translocation of the Northern Quoll* (Dasyurus hallucatus) *as a Conservation Response to the Spread of the Cane Toad* (Chaunus (Bufo) marinus) *in the Northern Territory, Australia*. Report submitted to the Natural Heritage Trust Strategic Reserve Program, as a component of project 2005/162: Monitoring and Management of Cane Toad Impact in the Northern Territory. Northern Territory Government of Australia, Darwin.

Rao NK, Hanson J, Dulloo ME, Ghosh K, Nowell D and Larinde M (2006) *Handbook for Genebanks No. 8: Manual for Seed Handling for Genebanks. No. 8*. Bioversity International, Rome.

Read JL (1999) A strategy for minimizing waterfowl deaths on toxic waterbodies. *Journal of Applied Ecology* **36**, 345–350.

Reese JG (1970) Reproduction in a Chesapeake Bay osprey population. *Auk* **87**, 747–759.

Reeve AF and Anderson SH (1993) Ineffectiveness of Swareflex reflectors at reducing deer vehicle collisions. *Wildlife Society Bulletin* **21**, 127–132.

Relyea RA (2005) The lethal impact of Roundup on aquatic and terrestrial amphibians. *Ecological Applications* **15**, 1118–1124.

Relyea RA (2006) The impact of insecticides and herbicides on the biodiversity and productivity of aquatic communities – response. *Ecological Applications* **16**, 2027–2034.

Rhodes LI (1972) Success of Osprey nest structures at Martin National Wildlife Refuge. *The Journal of Wildlife Management* **36**, 1296–1299.

Rhodes M (2002a) Update from the bat box project in Brisbane: a modified bat box design for Australian microbats. *The Australasian Bat Society Newsletter* **19**, 13–14.

Rhodes M (2002b) Criteria for successful bat houses. *The Australasian Bat Society Newsletter* **19**, 13–14.

Rhodes M and Jones D (2011) The use of bat boxes by insectivorous bats and other fauna in the greater Brisbane region. In *The Biology and Conservation of Australasian Bats*. (Eds B Law, P Eby, D Lunney and L Lumsden). Royal Zoological Society of NSW, Sydney (in press).

Rice EL (1977) Some roles of allelopathic compounds in plant communities. *Biochemical Systematics and Ecology* **5**, 201–206.

Rich C and Longcore T (2006) *Ecological Consequences of Artificial Night Lighting*. Island Press. Washington DC.

Robinson D (2006) Is revegetation in the Sheep Pen Creek area, Victoria, improving Grey-crowned Babbler habitat. *Ecological Management and Restoration* **7**, 93–104.

Robley A, Purdey D, Johnston M, Lindeman M, Busana F and Long K (2007) Experimental trials to determine effective fence designs for feral cat and fox exclusion. *Ecological Management and Restoration* **8**, 193–198.

Rodgers JA Jr and Smith HT (1995) Set-back distances to protect nesting bird colonies from human disturbance in Florida. *Conservation Biology* **9**, 89–99.

Rodriguez A and Rodriguez B (2009) Attraction of petrels to artificial lights in the Canary Islands: effects of the moon phase and age class. *Ibis* **151**, 299–310.

Rollan A, Real J, Bosch R, Tinto A and Hernandez-Matias A (2010) Modelling the risk of collision with powerlines in Bonelli's eagle Hieraaetus fasciatus and its conservation implications. *Bird Conservation International* **20**, 279–294.

Romanowski N (2009) *Planting Wetlands and Dams, Second Edition*. Landlinks Press, Melbourne.

Ronconi RA and St Clair CC (2006) Efficacy of a radar-activated on-demand system for deterring waterfowl from oil sands tailings ponds. *Journal of Applied Ecology* **43**, 111–119.

Rout TM and Hauser CE and Possingham HP (2009) Optimal adaptive management for the translocation of a threatened species. *Ecological Applications* **19**, 515–526.

Rowden P, Steinhardt D and Sheehan M (2008) Road crashes involving animals in Australia. *Accident Analysis and Prevention* **40**, 1865–1871.

Rowntree JK, Pressel S, Ramsay MM, Sabovljevic A, Sabovljevic M (2011) In vitro conservation of European bryophytes. *In Vitro Cellular and Developmental Biology - Plant*. **47**, 55–64.

Royal Botanical Gardens, Kew (2011) *The Millennium Seed Bank*, <http://www.kew.org/science-conservation/conservation-climate-change/millennium-seed-bank/index.htm>.

Royal Society for the Prevention of Cruelty to Animals (2008) *Royal Society for the Prevention of Cruelty to Animals 2008 Statistics – Wildlife Care*, <http://www.rspca-act.org.au/about-us/links-and-resources/>.

Rubolini D, Gustin M, Bogliani G and Garavaglia R (2005) Birds and power lines in Italy: an assessment. *Bird Conservation International* **15**, 131–145.

Ruiz-Jaen MC and Aide TM (2005) Restoration success: how is it being measured? *Restoration Ecology* **13**, 569–577.

Rusz PJ, Prince HH, Rusz RD and Dawson GA (1986) Bird collisions with transmission lines near a power plant cooling pond. *Wildlife Society Bulletin* **14**, 441–444.

Salmon M (2003) Artificial night lighting and sea turtles. *Biologist.* **50**, 163–168.

Salmon M (2006) Protecting sea turtles from artificial lighting at Florida's oceanic beaches. In *Ecological Consequences of Artificial Night Lighting.* (Eds C Rich and T Longcore) pp. 141–168. Island Press, Washington DC.

Samhouri JF, Levin PS and Ainsworth CH (2011) *Identifying Thresholds for Ecosystem-based Management,* <http://www.plosone.org/article/info%3Adoi%2F10.1371%2Fjournal.pone.0008907>.

Sander SM (1997) How to build a cave: a bold experiment in artificial habitat. *BATS* **15**, 8–11.

Sanderson KJ, Jaeger DA, Bonner JF and Jansen L (2006) Activity patterns of bats at house roosts near Adelaide. *Australian Mammalogy* **28**, 137–147.

Sands D (2008) Conserving the Richmond birdwing butterfly over two decades: Where to next? *Ecological Management and Restoration* **9**, 4–16.

Santos CD, Miranda AC, Granadeiro JP, Lourenco PM, Saraiva S and Palmeirim JM (2010) Effects of artificial illumination on the nocturnal foraging of waders. *Acta Oecologia* **36**, 166–172.

Saunders DA and Hobbs RJ (1991) The role of corridors in conservation: what do we know and where do we go? In *Nature Conservation 2: The Role of Corridors.* (Eds DA Saunders and RJ Hobbs) pp. 421–427. Surrey Beatty & Sons, Chipping Norton, UK.

Saunders DA and Ingram JA (1987) Factors affecting survival of breeding populations of Carnaby's cockatoo *Calyptorhynchus funereus latirostris* in remnants of native vegetation. In *Nature Conservation: The Role of Remnants of Native Vegetation.* (Eds DA Saunders, GW Arnold, AA Burbidge and JM Hopkins) pp. 250–258. Surrey Beatty & Sons, Chipping Norton, UK.

Saunders DA, Smith GT and Rowley I (1982) The availability and dimensions of tree hollows that provide nest sites for Cockatoos (Psittaciformes) in Western Australia. *Australian Wildlife Research* **9**, 541–556.

Seeding Victoria (2011) *Seeding Victoria Seed Banks.* Seeding Vitoria Inc., Creswick, <http://www.seedingvictoria.com.au/index.php>.

Shannon B (2007) Photopollution impacts on the nocturnal behaviour of the sugar glider (*Petaurus breviceps*). *Pacific Conservation Biology* **13**, 171–176.

Sharp T and Saunders G (2004) Development of a Model Code of Practice and Standard Operating Procedures for the Humane Capture, Handling or Destruction of Feral Animals in Australia. Final report for the Australian Government Department of the Environment and Heritage, Canberra.

Shaw JM, Jenkins AR, Smallie JJ and Ryan P (2010) Modelling power-line collision risk for the blue crane *Anthropoides paradiseus* in South Africa. *Ibis* **152**, 590–599.

Shepard DB, Kuhns, Dreslik MJ and Phillips CA (2008) Roads as barriers to animal movement in fragmented landscapes. *Animal Conservation* **11**, 288–296.

Shirely SM and Smith JN (2005) Bird community structure across riparian buffer strips of varying width in a coastal temperate forest. *Biological Conservation* **125**, 475–489.

Short J (2009) *The Characteristics and Success of Vertebrate Translocations Within Australia.* A final report to Department of Agriculture, Fisheries and Forestry, Canberra.

Short J and Smith A (1994) Mammal decline and recovery in Australia. *Journal of Mammalogy* **75**, 288–297.

Simberloff D and Cox J (1987) Consequences and costs of conservation corridors. *Conservation Biology* **1**, 63–71.

Simmons JM, Sunnucks P, Taylor AC and van der Ree R (2010) Beyond roadkill, radiotracking, recapture and Fst – a review of some genetic methods to improve understanding of the influence of roads on wildlife. *Ecology and Society* **15**, [online] <http://www.ecologyand-society.org/vol15/iss1/art9/>.

Sinclair Knight Merz (2007) *Traveston Crossing Dam Environmental Impact Statement.* Prepared for the Queensland Water Infrastructure Pty Ltd, Brisbane.

Sindel BM (2000) The history of integrated weed management. In *Australian Weed Management Systems.* (Ed. BM Sindel) pp. 253–266. RG and FJ Richardson. Meredith, Victoria.

Skyrail Rainforest Cableway (2011) *Skyrail Construction, Queensland,* <http://www.skyrail.com.au/construction.html>.

Smales I (2006) Impacts of Avian Collisions with Wind Power Turbines: An Overview of the Modelling of Cumulative Risks Posed by Multiple Wind Farms. Report for the Department of Environment and Heritage, Canberra.

Smales I, Muir S and Meredith C (2005) Modelled Cumulative Impacts on the Orange-bellied Parrot of Wind Farms Across the Species' Range in South-Eastern Australia. Report for the Department of Environment and Heritage, Canberra.

Smith D (2011) *Dick Smith's Population Crisis.* Allen and Unwin, Sydney.

Smith GC and Agnew G (2002) The value of 'bat boxes' for attracting hollow-dependent fauna to farm forestry plantations in southeast Queensland. *Ecological Management and Restoration* **3**, 37–46.

Smith GC and Carlile N (1993) Methods for population control within a silver gull colony. *Wildlife Research* **20**, 219–226.

Soderquist TR, Traill BJ, Faris F and Beasley K (1996) Using nest boxes to survey for the brush-tailed phascogale *Phascogale tapoatafa. The Victorian Naturalist* **113**, 256–261.

Soldatini C, Albores-Barajas YV, Torricelli P and Mainardi D (2008) Testing the efficacy of deterring systems in two gull species. *Applied Animal Behaviour Science* **110**, 330–340.

Soule ME and Gilpin ME (1991) The theory of wildlife corridor capability. In *Nature Conservation 2: The Role of Corridors.* (Eds DA Saunders and RJ Hobbs) pp. 3–8. Surrey Beatty & Sons, Chipping Norton, UK.

Souter NJ, Bull CM and Hutchinson MN (2004) Adding burrows to enhance a population of the endangered pygmy blue tongue lizard, *Tiliqua adelaidensis. Biological Conservation* **116**, 403–408.

Spanier E (1980) The use of distress calls to repel night herons (*Nyticorax nyticorax*) from Fish Ponds. *Journal of Applied Ecology* **17**, 287–294.

Spooner P, Lunt I and Robinson W (2002) Is fencing enough? The short-term effects of stock exclusion in remnant grassy woodlands in southern New South Wales. *Ecological Management and Restoration* **3**, 117–126.

Statham M (2009) *Wallaby Proof Fencing: A Planning Guide for Tasmanian Primary Producers.* Tasmanian Institute of Agricultural Research, Hobart.

Steele WK (2001) Factors influencing the incidence of bird-strikes at Melbourne Airport 1986–2000. In *Bird Strike Committee Proceedings 2001 Bird Strike Committee-USA/Canada, Third Joint Annual Meeting,* 27–30 August, Calgary, Canada. pp. 144–157. Bird Strike Committee, University of Nebraska, Lincoln.

Stevens GR, Rogue J, Weber R and Clark L (2000) Evaluation of a radar-activated, demand-performance bird hazing system. *International Biodeterioration and Biodegradation* **45**, 129–137.

Stockwell MP, Clulow S, Clulow J and Mahony M (2008) The impact of the Amphibian Chytrid Fungus *Batrachochytrium dendrobatidis* on a green and golden bell frog *Litoria aurea* reintroduction program at the Hunter Wetlands Centre Australia in the Hunter Region of New South Wales. *Australian Zoologist* **34**, 379–386.

Stone EL, Jones G and Harris S (2009) Street lighting disturbs commuting bats. *Current Biology* **19**, 1123–1127.

Swinburn ML, Fleming PA, Criag MD, Grigg AH, Garkaklis MJ, Hobbs RJ and Hardy GS (2007) The importance of grasstrees (*Xanthorrhoea preissii*) as habitat for mardo (*Antechinus flavipes leucogaster*) during post fire recovery. *Wildlife Research* **34**, 640–651.

Taylor BD and Goldingay RL (2003) Cutting the carnage: wildlife usage of road culverts in north-eastern New South Wales. *Wildlife Research* **30**, 529–537.

Taylor BD and Goldingay RL (2009) Can road-crossing structures improve population viability of an urban gliding mammal? *Ecology and Society* **14**, 13.

Taylor BD and Goldingay RL (2010) Roads and wildlife: impacts, mitigation and implications for wildlife management in Australia. *Wildlife Research* **37**, 320–331.

Thierry A, Lettink M, Besson AA and Cree A (2009) Thermal properties of artificial refuges and their implications for retreat-site selection in lizards. *Applied Herpetology* **6**, 307–326.

Thomson B (2002) Australian Handbook for the Conservation of Bats in Mines and Artificial Cave-Bat Habitats. Australian Centre for Mining Environmental Research, Brisbane.

Tolga Bat Hospital (2010) *Wildlife Friendly Fencing*, Tolga Bat Hospital, Atherton, Queensland, <http://www.wildlifefriendlyfencing.com/>.

Tolhurst KG, Flinn DW, Loyn RH, Wilson AAG and Foletta I (1992) Ecological Effects of Fuel Reduction Burning in a Dry Sclerophyll Forest – A Summary of Principle Research Findings and their Management Implications. Department of Conservation and Environment, Melbourne.

Tongway DJ and Ludwig JA (1990) Vegetation and soil patterning in semi-arid mulga lands of eastern Australia. *Australian Journal of Ecology* **15**, 23–34.

Tongway DJ and Ludwig JA (2010) Restoring Disturbed Landscapes Putting Principles into Practice. CSIRO Publishing, Melbourne.

Trainor R (1995) *Artificial Nest-hollows*. Report on the use of artificial hollows in the box-ironbark forest surrounding Maryborough, Central Victoria 1982-1995. *The Bird Observer* **759**, 5–7.

Transport Canada (1998) *Evaluation of the Efficacy of Products and Techniques for Airport Bird Control (03/1998).* Transport Canada Safety and Security Aerodrome Safety Branch, Canada.

Transport Canada (2002) *Wildlife Control Procedures Manual.* Transport Canada Safety and Security Aerodrome Safety Branch, Canada.

Triggs B (2009) *Wombats, Section Edition.* CSIRO Publishing, Melbourne.

Trombulak SC and Frissell CA (2000) Review of ecological effects of roads on terrestrial and aquatic communities. *Conservation Biology* **14**, 18–30.

Tuttle MD, Kiser M and Kiser S (2004) *The Bat House Builder's Handbook.* Bat Conservation International, Austin, Texas.

Uhlmann L (2010) 'Koalas return to rehabilitated mine sites'. Bayside Bulletin, Brisbane, <http://www.baysidebulletin.com.au/news/local/news/general/koalas-return-to-rehabilitated-mine-sites/2021600.aspx?src=rss>.

Underwood JG (2011) Combining landscape-level conservation planning and biodiversity offset programs: a case study. *Environmental Management* **47**,121–129.

Ujvari M, Baagoe HJ and Madsen AB (1998) Effectiveness of wildlife warning reflectors in reducing deer-vehicle collisions: A behavioral study. *Journal of Wildlife Management* **62**, 1094–1099.

Ujvari M, Baagoe HJ and Madsen AB (2004) Effectiveness of acoustic road markings in reducing deer-vehicle collisions: a behavioural study. *Wildlife Biology* **10**, 155–159.

Valladares-Padua C, Cullen L Jr and Padua S (1995) A pole bridge to avoid primate road kills. *Neotropical Primates* **3**, 13–15.

Vallee L, Hogbin T, Monks L, Makinson B, Matthes and Rossetto M (2004) *Guidelines for the Translocation of Threatened Plants in Australia, Second Edition*. Australian Network for Plant Conservation, Canberra.

van der Ree R (1999) Barbed wire fencing as a hazard for wildlife. *The Victorian Naturalist* **116**, 210–217.

van der Ree R and Bennett A (2003) Home range of the squirrel glider (*Petaurus norfolcensis*) in a network of remnant linear habitats, *Journal of Zoology* **259**, 327–336.

van der Ree R, Bennett AF and Gilmore DC (2003) Gap-crossing by gliding marsupials: thresholds for use of isolated woodland patches in an agricultural landscape. *Biological Conservation* **115**, 241–249.

van der Ree R, Gulle N, Holland K, van der Grift E, Mata C and Suarez F (2007) Overcoming the barrier effect of roads – how effective are mitigation strategies? In *Proceedings of the 2007 International Conference on Ecology and Transportation*. 20–25 May 2007 Little Rock, Arkansas (Eds CL Irwin, D Nelson and KP McDermott) pp. 423–431. Center for Transportation and the Environment, North Carolina State University, Raleigh, NC.

van der Ree R, Clarkson DT, Holland K, Gulle N and Budden M (2008a) *Review of Mitigation Measures Used to Deal with the Issue Of Habitat Fragmentation by Major Linear Infrastructure*. Report for Department of Environment, Water, Heritage and the Arts (DEWHA) Contract No. 025/2006. DEWHA, Canberra.

van der Ree R, Soanes K, Cesarini S, McCall S, Taylor A, Sunnucks P, Harper M and Gulle N (2008b) Rope bridge story revisited. Episode 30 of The University of Melbourne 'Visions' Video podcasts, <http://visions.unimelb.edu.au/episode/41>.

van der Ree R, Heinze D, McCarthy M and Mansergh I (2009) Wildlife tunnel enhances population viability. *Ecology and Society* **14**, 7 [online]. < www.ecologyandsociety.org/vol14/iss2/art7/ES-2009-2957.pdf >.

van der Ree R, Cesarini S, Sunnucks P, Moore JL and Taylor A (2010). Large gaps in canopy reduce road crossing by a gliding mammal. *Ecology and Society*, **15** [online]. < www.ecologyandsociety.org/vol15/iss4/art35/ES-2010-3759.pdf >.

Van Dyck S and Strahan R (2008) *The Mammals of Australia, Third Edition*. Reed New Holland, Sydney.

van Langevelde F and Jaarsma CF (2009) Modelling the effect of traffic calming on local animal population persistence. *Ecology and Society* **14**, [online] < www.ecologyandsociety.org/vol14/iss2/art39/ES-2009-3061.pdf >.

van Slageren MW (2003) The Millennium Seed Bank: building partnerships in arid regions for the conservation of wild species. *Journal of Arid Environments* **54**, 195–201.

van Tets GF, Vestjens WJM and Slater E (1969) Orange runway lighting as a method for reducing bird strike damage to aircraft. *Wildlife Research* **14**, 129–151.

Veage L and Jones DN (2007) *Breaking the Barrier: Assessing the Value of Fauna-friendly Crossing Structures at Compton Road*. Report for Brisbane City Council. Centre of Innovative Conservation Strategies, Griffith University, Brisbane.

Villarroya A and Puig J (2010) Initial steps in the design of compensation measures for habitat and landscape effects of road construction. In *2010 IENE International Conference on Ecology and Transportation: Improving Connections in a Changing Environment*. 27 September–1 October 2010. pp. 44–47. Infra Eco Network Europe, Velence, Hungary.

Vitelli JS and Pitt JL (2006) Assessment of current weed control methods relevant to the management of the biodiversity of Australian rangelands. *The Rangeland Journal* **28**, 37–46.

Voigt M (2006) *Establishing Frog Habitats on Your Property*. Frogfacts No.3. FATS Group, Sydney.

Walker KF (2008) Environmental Impact Statement for Traveston Crossing Dam (Mary River, Queensland): A Review with Regard for Species of Concern under the EPBC Act 1999. Report to the Department of Environment, Water, Heritage and the Arts, Canberra.

Wallach AD and O'Neill AJ (2009) Artificial water points: hotspots of extinction or biodiversity? *Biodiversity Conservation* **142**, 1253–1254.

Waring GH, Griffis JL and Vaughn ME (1991) White-tailed deer roadside behaviour, wildlife warning reflectors, and highway mortality. *Applied Animal Behaviour Science* **29**, 215–223.

Watson J, Watson A, Paull D and Freudenberger D (2003) Woodland fragmentation is causing the decline of species and functional groups of birds in south-eastern Australia. *Pacific Conservation Biology* **8**, 261–270.

Webb JK and Shine R (2000) Paving the way for habitat restoration: can artificial rocks restore degraded habitats of endangered reptiles? *Biological Conservation* **92**, 93–99.

Weston MA, Antos MJ and Glover HK (2009) Birds, buffers and bicycles: a review and case study of wetland buffers. *The Victorian Naturalist* **126**, 76–86.

Weston NG (2003) The provision of canopy bridges to reduce the effects of linear barriers on arboreal mammals in the Wet Tropics of northeastern Queensland. MSc thesis. James Cook University, Cairns.

White AW (2006) A trail using salt to protect green and golden bell frogs from chytrid infection. *Herpetofauna* **36**, 93–96.

White AW (2008) The 2009 Sydney 400 V8 Supercar Event, Sydney Olympic Park, 4-6 December 2009 Green And Golden Bell Frog (*Litoria aurea*) Impact Assessment. Prepared For V8 Supercars Australia. Southport, Queensland.

White AW and Pyke GH (2008) Green and golden bell frogs in New South Wales: current status and future prospects. *Australian Zoologist* **34**, 319–333.

White GH, Morozow O, Allan JG and Bacon CA (1996) Minimisation of impact during exploration. In *Environmental Management in the Australian Minerals and Energy Industries: Principles and Practices*. (Ed. D Mulligan) pp. 99–130. University of New South Wales Press in association with the Australian Minerals and Energy Environmental Foundation, Sydney.

White LP (1971) Vegetation strips on sheet wash surfaces. *The Journal of Ecology* **59**, 615–622.

Wilkins S, Keith DA and Adam P (2003) Measuring success: evaluating the restoration of grassy eucalypt woodland on the Cumberland Plain, Sydney, Australia. *Restoration Ecology* **11**, 489–503.

Williams B (2011) Fencing in cane toads could stop breeding boom. *Courier Mail* 24 February 2011.

Williams J (2000) Managing the Bush: Recent Research Findings from the EA/LWRRDC National Remnant Vegetation Research and Development Program. Land and Water Resources Research and Development Corporation, Canberra.

Wilson RF, Marsh H and Winter J (2007) Importance of canopy connectivity for home range and movements of the rainforest arboreal ringtail possum (*Hemibelideus lemuroides*). *Wildlife Research* **34**, 177–184.

Winning G and Baharrell M (1998) Design of habitat wetlands, wetland rehabilitation. In *The Constructed Wetlands Manual*. (Ed. New South Wales Department of Land and Water Conservation). New South Wales Department of Land and Water Conservation, Sydney.

Winter JW, Moore NJ and Wilson RF (2008) Lemuroid Ringtail Possum. In *The Mammals of Australia, Third Edition*. (Eds S Van Dyck and R Strahan) pp. 238–240. Reed New Holland, Sydney.

Winter JW and Moore NJ (2008) Herbert River Ringtail Possum. In *The Mammals of Australia, Third Edition.* (Eds S Van Dyck and R Strahan) pp. 250–252. Reed New Holland, Sydney.

Witheridge G (2010) Erosion and Sediment Control – A Field Guide for Construction Site Managers – Version 2. Catchments and Creeks Pty Ltd, Brisbane.

Wolff JO, Fox T, Skillen RR and Wang GM (1999) The effects of supplemental perch sites on avian predation and demography of vole populations. *Canadian Journal of Zoology* **77**, 535–541.

Woltz HW, Gibbs JP and Ducey PK (2008) Road crossing structures for amphibians and reptiles: Informing design through behavioural analysis. *Biological Conservation* **141**, 2745–2750.

Xstrata (2008) *Xstrata Partners with Queensland Environment Protection Authority to Save Endangered Wombat*, Xstrata Copper, Brisbane, <http://www.xstrata.com/media/news/2008/05/09/1530CET/>.

Index